Underwater
Labriform-Swimming Robot

Topics in Systems Engineering

Print ISSN: 2810-9090
Online ISSN: 2810-9104

Series Editors: Luigi Fortuna *(Università degli Studi di Catania, Italy)*
Arturo Buscarino *(Università degli Studi di Catania, Italy)*

The Series aims to cover a wide spectrum of Engineering topics but with a strong characteristic of interdisciplinarity using less technical items. It could be a series including thinking aspects, history topics linked with engineering topics from civil engineering to industrial engineering and so on. The contributions will cover a wide series of engineering topics and could be useful both for educational purposes and for research information. The idea is to combine technology, science, arts and social science with some emerging subjects.

Published:

Vol. 1:   *Underwater Labriform-Swimming Robot*
by Farah Abbas Naser, Mofeed Turky Rashid and Luigi Fortuna

Topics in Systems Engineering — Volume 1

Underwater
Labriform-Swimming Robot

Farah Abbas Naser
University of Basrah, Iraq

Mofeed Turky Rashid
University of Basrah, Iraq

Luigi Fortuna
University of Catania, Italy

World Scientific

NEW JERSEY · LONDON · SINGAPORE · BEIJING · SHANGHAI · HONG KONG · TAIPEI · CHENNAI · TOKYO

Published by

World Scientific Publishing Co. Pte. Ltd.

5 Toh Tuck Link, Singapore 596224

USA office: 27 Warren Street, Suite 401-402, Hackensack, NJ 07601

UK office: 57 Shelton Street, Covent Garden, London WC2H 9HE

Library of Congress Control Number: 2021049384

British Library Cataloguing-in-Publication Data
A catalogue record for this book is available from the British Library.

Topics in Systems Engineering — Vol. 1
UNDERWATER LABRIFORM-SWIMMING ROBOT

ISBN 978-981-123-739-3 (hardcover)
ISBN 978-981-123-740-9 (ebook for institutions)
ISBN 978-981-123-741-6 (ebook for individuals)

For any available supplementary material, please visit
https://www.worldscientific.com/worldscibooks/10.1142/12292#t=suppl

Typeset by Stallion Press
Email: enquiries@stallionpress.com

To the spirt of my beloved father...
To the candle of my life Mum, brothers and sisters...
To the greatest gift that I ever had in my life, my husband and
our lovely children, Fatima, Abdullah and Maryam....
Thanks all for your unconditional love and support....

Farah Abbas Naser

To my family
parents, wife and sons
To my brothers, sisters and friends

Mofeed Turky Rashid

To the friendship

Luigi Fortuna

PREFACE

Recently, the applications of the swimming robot have widely increased. Due to the urgent need to use it in operations where human intervention is difficult, such as the exploration of the deep sea, military intervention, and entering areas with high water pollution, that threatens the lives of divers, the importance of using swimming robots as an alternative to humans in the underwater application has gained.

Currently, the swimming modes of underwater organisms have been widely used in various designs for swimming robots with the excellence of being high maneuverable, approximately real, and efficient. While organisms are swimming underwater, there is an interaction between the morphology of a fish body, its fins, and the environment to achieve high performance of motion. These features of organisms make the robot's design, development, and control a difficult challenge. High maneuverability and efficiency are some of the potential benefits of swimming robots compared to traditional propeller-actuated underwater vehicles. Therefore, this book will present the design and implementation of a swimming robot based on a Labriform swimming mode and highlights several issues that constitute the important factors in the success of the design while swimming underwater.

It begins by studying the effect of the morphology or shape of the pectoral fins on swimming performance. The analytical study of the effect of fin shape on the performance of swimming robots is a vital contribution to the robotic community. Three different concave pectoral fins named Fin1, Fin2, and Fin3 with different surface areas are carefully examined. The results of the experiments showed that Fin1 is best suited for diving because it produces a higher thrust than the others, while Fin2 is used to carry out forward movement and steering operations, and Fin3 is excluded from this work because it is ineffective for robot acceleration.

Second is the design and implementation of a swimming robot propelled by two pectoral fins, where the full-body design of the swimming robot is presented. Further, studying the case of achieving highest swimming speed by changing the ratios (R_{F}) between the power stroke velocity and the recovery stroke velocity, the optimum R_{F} ratio is obtained at 3:1.

Third, two studies have been presented to achieve the orientation process. These two strategies include studying the ability of the rigid body of boxfish to achieve high performance in the guiding process based on the caudal fin, while the other study includes modeling and experimental validation of this model to achieve the steering process by the pectoral fins. The thrust force in these two systems is generated by pectoral fins.

Finally, the design and modeling of a swimming robot that can realize the diving process based on the sliding block mechanism has been studied. Two swimming robots designs have been presented, in which, the propulsion of the first swimming robot is based on the Carangiform mechanism in swimming, the second is based on the principle of the Labriform mechanism, while the diving system in these two models is based on Controlling the Center of Gravity (CoG) using a sliding block system.

The kinematics and dynamics model of the swimming robot have been studied for all these mentioned cases, while an evaluation of the total hydrodynamic forces exerted on the swimming robot's body is studied via the computational fluid dynamics (CFD) method from SOLIDWORKS®. The design has been validated theoretically by MATLAB and examined practically by implementing several experiments.

Mofeed Turky Rashid and Luigi Fortuna

CONTENTS

Chapter 1

INTRODUCTION

1.1. Overview

During the last decades, a huge revolution has occurred in the fields
of robotics, especially with regard to their types and environment. In
recent years, swimming robots have developed greatly due to their
many important applications. Underwater tasks are usually difficult
for human capabilities, as the diver needs to remain underwater
for long periods without oxygen, while the use of devices and
equipment that provide oxygen to the diver causes an obstacle to
his movement underwater and affects the performance of the task
completion. Other difficulties that hinder and limit the work of divers
are water pressure, as divers can dive to limited depths. Further,
the visibility underwater is a difficult challenge, as the density of
water and the impurities present in it reduce the extent of vision.
There are important determinants that must be taken seriously,
which has led to resorting to swimming robots due to the risks
that humans face underwater due to natural phenomena (storms,
marine eddies, ice, lava, etc.) or as a result of predatory and toxic
organisms (whales, sharks, poisonous invertebrates, etc.). Due to
these aforementioned issues, it has become imperative to resort to
underwater swimmers doing the work instead of humans. So in 1989,
Massachusetts Institute of Technology first published research that
includes the idea of designing fish robots.

In this paragraph, we will try to give an appropriate definition of
the types of robots that swim in the water. There are three common
names for them, which are swimming robots, underwater robots, and

fish robots. Swimming robot indicates robots that swim above water or underwater, in which these robots can achieve two-dimensional motion by moving above the water that only include two degrees of freedom (2 DOF), or achieving three-dimensional motion by moving underwater, diving inside water with this motion includes 3 DOF. The proposed design of such robots can be inspired by the designer himself based on different phenomena or rules in the real-world or by ideas inspired by the organisms in real-world. Further, several researchers have proposed swimming robot systems by observing the organisms in the real world whether it is swimming over water or underwater. Figure 1.1, shows several types of swimming robots.

An underwater robot is another type of robot that only swims inside the water, moves in a three-dimensional space, and achieves the diving process. There are many methods used in designing such a robot that depend on the mechanism of sailing vehicles. In addition, there are several types of this robot associated with the size of the robot, namely, a Nano robot, a macro robot, and an underwater robot. This type of robot is the most widespread as it is used in many applications such as navigation, scientific and military, further there are many studies to use it in health applications depending on the development of the Nano robot. The underwater robot is currently available in the market and is ready for use in different applications. Similar to the swimming robot type, some underwater robot designs are proposed, inspired by the design from swimming organisms, Artemia, Sperm, etc. (See Figure 1.2).

The fish robot is the robot inspired by fish with respect to body structure and swimming mechanism. There are many studies that deals with this type of robot, which has rapidly evolved during this decade from the viewpoint of performance and stability. These robots are expected to be introduced in many important applications, including military, scientific and environmental, as these robots emulate the movement of real fish and may appear to be one of them, and this leads to saving living organisms in the event of natural disasters. On the other hand, it can be used to attract fish during scientific studies or to observe underwater organisms. Experiments of this type of robot have shown that their behavior seems similar

(a) The swimming robot "Elisabeth"

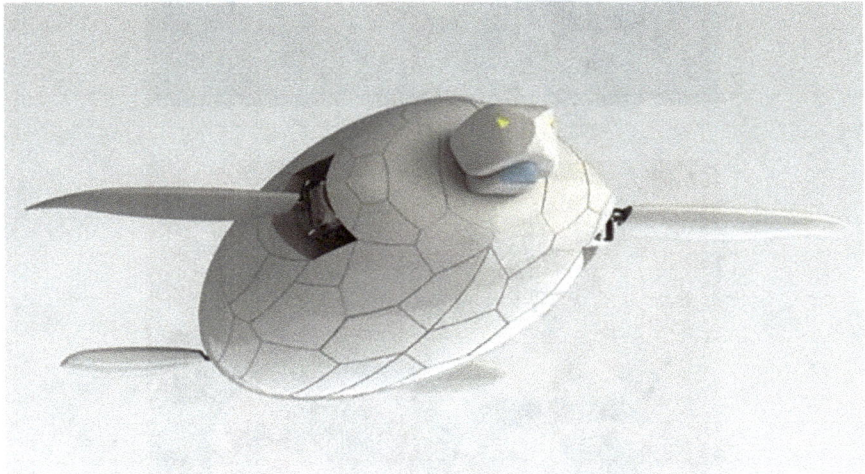

(b) A Sea Turtle Robot

Figure 1.1. Samples of Swimming Robot.

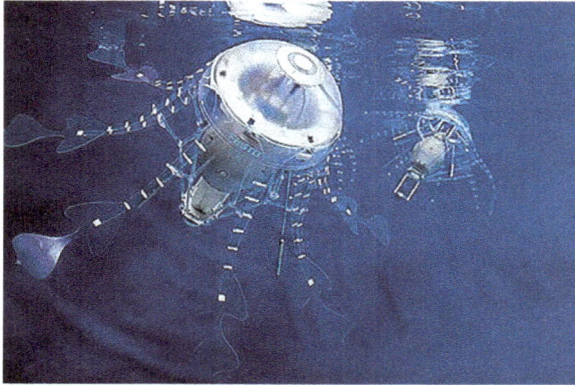

(a) Robotic jellyfish

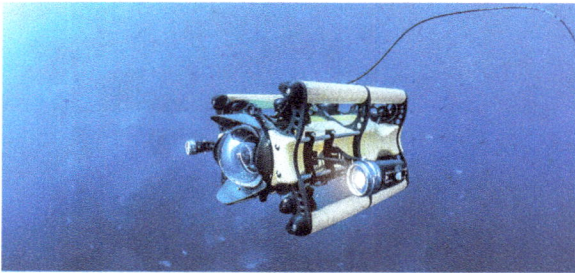

(b) Underwater robot

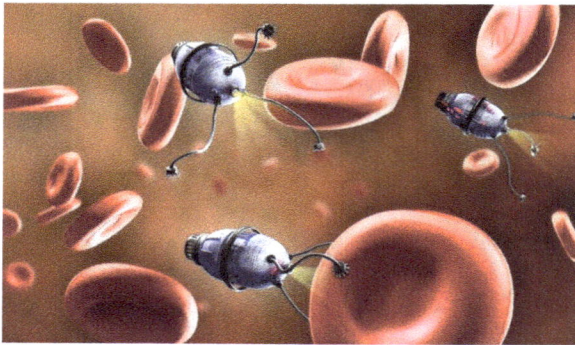

(c) Swimmer Micro robot will prevent diseases

Figure 1.2. Samples of underwater robot.

(a) A robot fish will 'swim' in Asturias in search of contamination

(b) The Airacuda fish robot.

Figure 1.3. Samples of the fish robot.

to real fish and it is difficult to distinguish between the movement of the real fish and the robot. In addition, this type of robot has shown good performance during diving and maneuvering, and this facilitates exploration operations or bypassing difficult environmental conditions (see Figure 1.3).

As mentioned above, swimming robots can be classified according to their ability to swim; therefore, there are several designs for swimming robot that differ from each other. This difference results from three basic parts in the robot that are mainly responsible for swimming underwater or on the surface of the water. Namely the external design of the body and how it resists the water flow, the actuator systems generating the forward thrust, which is needed for swimming, and the diving systems that the robot uses for diving and buoyancy.

There are several types of swimming robot platforms, most of them are inspired by the body shape of organisms from nano-size to big size. Some of these designs have been modified to be appropriate for the applications, and others seem similar to the organisms such as Cells, Insects, Crustaceans, Reptiles, Amphibians, and Fish. Figure 1.4 shows several designs of the swimming robot's body inspired by the organisms.

As known, several methods have been used for generating the thrust force in the swimming robot. Some of these methods depend on turbines, while others depend on fins, which are inspired by organisms. For fins, there are several types, depending on the organisms and the method of swimming. Therefore, the organisms can be categorized according to their fins as the Body/Caudal Fin (BCF), and Median/Paired Fin (MPF) swimming modes as shown in Figure 1.5. Furthermore, according to the motion generated from the fin type, they can be classified into the undulatory-motion type and the oscillatory-motion type.

In the BCF of undulatory-motion type, the motion created here resembles a wave structure produced from the fish head to its tail, for example, Carangiform, and Subcarangiform modes. For oscillatory-motion type, the movement is created by turning the body or caudal fin to propel the fish, for example, Thunniform and Ostraciiform modes. The other method of movement in swimming is the MPF-undulatory Gymnotiform, Amiiform, and Balastiform motion where oscillatory-type of movement includes Labriform and Tetraodmti-form. Another type of MPF that depends both on undulatory and oscillatory locomotion is known as Rajiform. About 15% of the fish

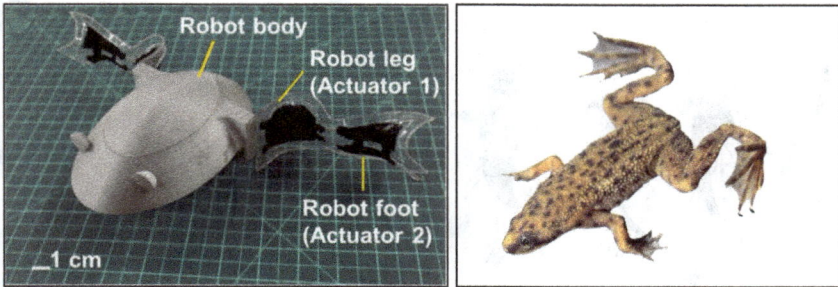

(a) Frog like robot.

(b) Snake like robot.

(c) Salamander like robot.

(d) Fish like robot.

Figure 1.4. Several types of swimming robot body design.

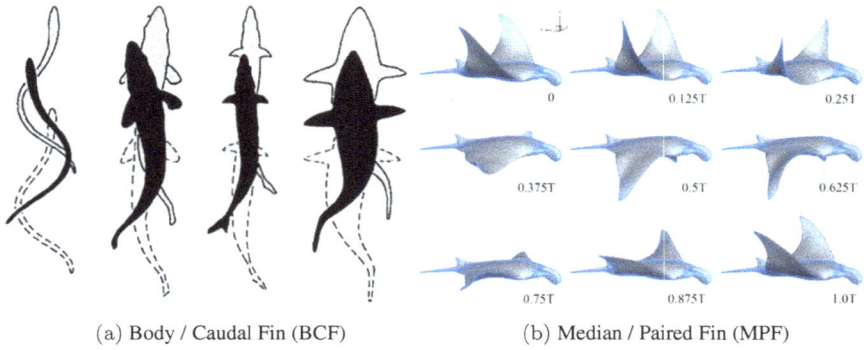

(a) Body / Caudal Fin (BCF)   (b) Median / Paired Fin (MPF)

Figure 1.5. Swimming modes.

(a) Anatomy of the pectoral fin in a highly maneuverable coral reef fish, the parrotfish.

(b) Porcupine fish swim by undulating their pectoral fins.

Figure 1.6. Fish characterization according to steering process.

depends on MPF modes as the main propulsion mechanism, while there are very large numbers based on BCF modes for propulsion and use MPF modes for stabilization and maneuvering purposes.

There is another important issue related to bionic robot classification, which is the steering process that should facilitate a change in the direction of the swimming robot. The steering process achieves the robot orientation in the two-dimensional space, inspired by fish that achieve it with the help of the caudal fin or by the pectoral fins. Therefore, the fish achieves thrust force by caudal fin and steering process by pectoral fins. The second type of fish achieves thrust force by pectoral fins and steering process by caudal fin. The third type of fish achieves thrust force and steering process by caudal fin and pectoral fins are used for balancing only. Finally, some fish achieve thrust force and steering process by pectoral fins (see Figure 1.6).

While swimming in three-dimensional space, a swimming robot needs a system called a diving system that is used to achieve diving and buoyancy operations. There are several methods of diving that swimming robots can achieve, some of which are achieved by pectoral fins and caudal fin, and others are done by caudal fin and center of gravity concepts as shown in Figure 1.7.

The different modes of swimming mechanism in organisms are used in developing various designs for swimming robots with the excellence of being high maneuverable, approximately real, and efficient. There is an interaction between the shape of a fish body, its fins, and the environment in which it swims. These features make the robot's design, development, and control an important challenge. High maneuverability and efficiency are some of the potential benefits of swimming robots in comparison to the traditional propeller-actuated underwater vehicles.

1.2. Literature Review

As mentioned above the fish swimming in several modes achieves high swimming performance and excellent maneuverability. This has led researchers to study fish swimming modes deeply, which started in the biological fields by observing fish in water environments and

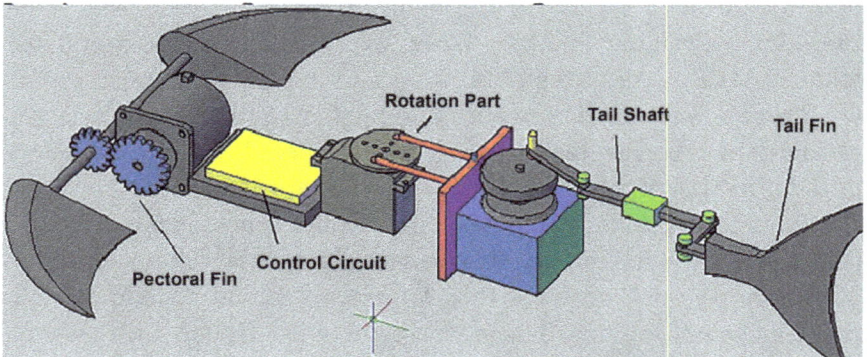

(a) Diving process is achieved by pectoral fins.

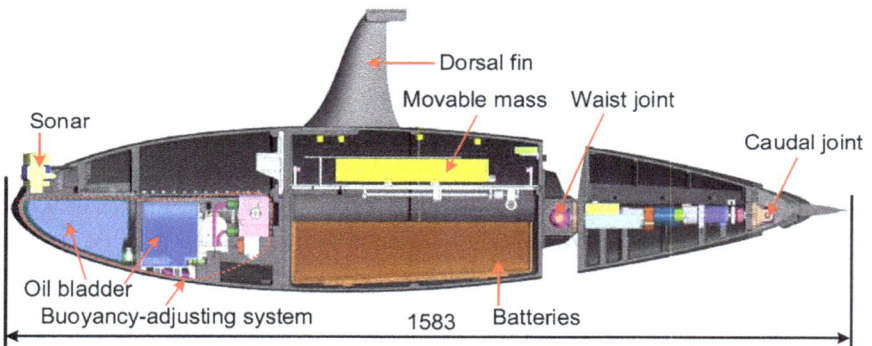

(b) Diving process is achieved by changing center of gravity.

Figure 1.7. Some types of diving system in swimming robot.

then modeled physically by physicists, these physical models have been simplified by mathematicians into mathematical models known as kinematic models in order to be suitable for computer application. Nowadays, these studies are widely employed to comprehend fish locomotion that is generated from the movement of fins and/or body; literally, there are many studies concerned with designing underwater swimming robots in detail, with various levels of swimming complexity and modes. Swimming robots can generate forward thrust force in multiple ways; it can use caudal fin (tail fin), pectoral fins or use both.

In light of the perceptions inspired by nature, various types of actuators have been used to enhance maneuverability and to increase the efficiency of designing swimming robots. The most common form of underwater actuation tool is the use of motors. These motors may require either a single or a multiple number of joints connected by rigid links to the tail fin or the pectoral fin. Another form of actuation system is the use of smart material.

In the last annuals, the impact of pectoral fins in the swimming process has appeared, for this reason, the design of underwater swimming robots that employed pectoral fins for generating thrust force was deeply studied. Most of these researches have talked about the importance of pectoral fins in the maneuverability and stability of robotic fish. Morphology, kinematics, and hydrodynamics have gained the largest attention in studies in terms of the analytical and computational fluid dynamics (CFD) aspects. The researchers have demonstrated the pectoral fin propulsion and movement of a live fish, understanding the calculations of hydrodynamic forces engaged with pectoral fin actuation. There are several studies in the morphology of the pectoral fin, considering the shape of the fin with one degree of freedom (DOF). Some of these swimming robots are shown in Figure 1.8.

The observation and understanding of the control of flexible fins of fishes is the fundamental resource for modeling and designing actuators to perform similar operations of these fins, which are employed by swimming robots for generating the thrust force. In 2010, a group of researchers from the USA developed a sensory model inspired by the pectoral fins of sunfish that can improve the motion and maneuver performance of a swimming robot. This pectoral fin is a boirobitic model that includes a sensorimotor system evolved by investigating the relation between sensory information, fin motion, and thrust force. The proposed pectoral fin generates the forces and motions of the biological fin during the swimming and steering process, which includes a set of small sensors in a lateral line and the receptors that should be within the pectoral fin. Several experiments have achieved validation of the proposed design, where the results support the success of this system.

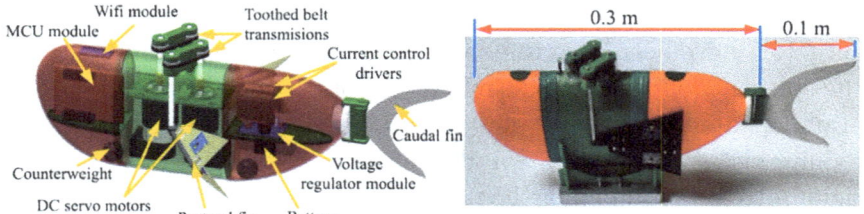

Wifi module
MCU module
Toothed belt transmisions
Current control drivers
Caudal fin
Counterweight
Voltage regulator module
DC servo motors
Pectoral fin
Battery

0.3 m   0.1 m

(a) Folding fin robot.

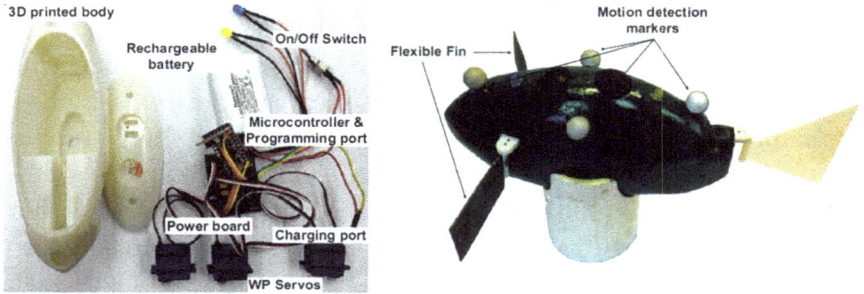

3D printed body
Rechargeable battery
On/Off Switch
Microcontroller & Programming port
Power board
Charging port
WP Servos
Flexible Fin
Motion detection markers

(b) Flexible pectoral fin.

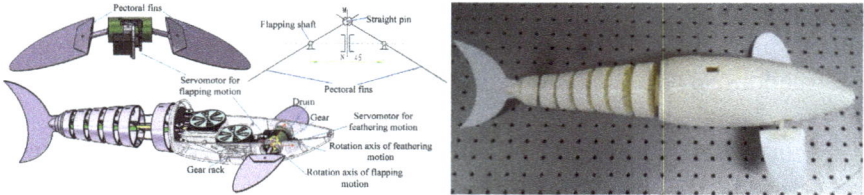

Pectoral fins
Flapping shaft
Straight pin
Servomotor for flapping motion
Pectoral fins
Drum
Gear
Servomotor for feathering motion
Rotation axis of feathering motion
Gear rack
Rotation axis of flapping motion

(c) 2-DOF robot.

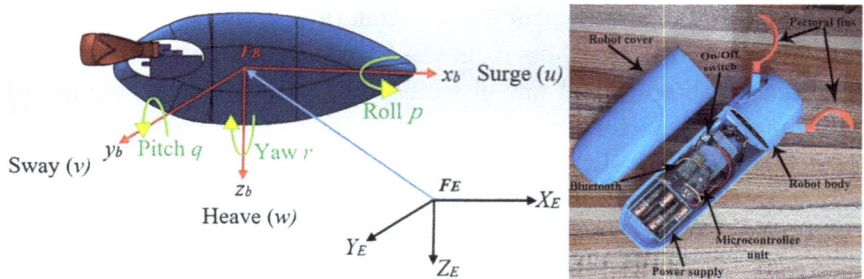

x_b Surge (u)
Roll p
y_b Pitch q Yaw r
Sway (v)
z_b
Heave (w)
F_E
X_E
Y_E
Z_E

Robot cover
On/Off switch
Pectoral fins
Bluetooth
Robot body
Microcontroller unit
Power supply

(d) Concave pectoral fin.

Figure 1.8. Examples of reported robotic fish prototypes with paired pectoral fins.

In 2014, authors from Peking University, China proposed robotic fish that depends on central pattern generator (CPG) in the motion control system. This model uses both pectoral fins and caudal fin as a main propulsion system. The pectoral fins are designed to rotate 360°, while the caudal fin oscillates at 120°. In this manner, the water flow produces a reactive force due to the momentum conservation and pushes the robotic body in a direction parallel to the oscillating fins. Therefore, the robot can achieve different movement patterns like swimming forward, backward, turning, rolling, pitching, and breaking.

In 2015, some researchers from Beihang University, Beijing, China have proposed a swimming robot like Cownose. The idea of this robot is inspired by the observation of Cownose swimming. The swimming robot uses two bionic pectoral fins for generating thrust force, and the motions of these fins are the combination of oscillating motion and chordwise twisting motion. The fins motion is inspired by analyzing the motion of the Cownose ray's pectoral fins. Further, the researchers designed several experiments to validate the proposed swimming robot, in which the results show the ability of this robot to swim in the pool that emulates reality. Further, the experiments showed that the pectoral fins can achieve a maximum forward speed of 0.43 m/s (0.94 times of body length per second) and excellent steering at a minimum radius.

Fishes have high efficiency and high maneuverability underwater, in which their fins have evolved over many generations to adapt their propelling in a complex environment. On the other hand, insects have wings developed over many years to achieve high-performance flying and steering in 3D space. Therefore, in 2016, a group of researchers working at the University of Science and Technology of China have proposed and developed an integrative biometric robotic fish depending on the features inspired by the high performance of swimming of fish and flying of insects. This robot has a couple of caudal fins in parallel and placed at the tail of the robotic fish which flap in the opposite directions that lead to generating opposing lateral forces, therefore propulsion forward will be generated to achieve high performance and stable swimming. In addition, two pectoral fins have been designed like wings of an insect, which are used to enhance the performance of swimming and maneuverability in the vertical plane. Moreover, the design of the swimming robot has a CPG model to perform several maneuvering motions, autonomous obstacle avoidance, and motion switching in 3D space. The proposed design has been validated by several experiments, where the results of these experiments show that the swimming robot can achieve high-performance swimming and maneuvering.

The Manta ray is the most suitable organism for inspired swimming robots due to the high performance flapping pectoral fin that has smoothness motion, high payload capacity, high stability, and good maneuverability. Therefore, in Norway, during 2016, a group of researchers developed a bioinspired swimming robot based on a comprehensive analysis of the structure of Manta ray, and they found there is a relationship between the size of pectoral fins and the shape of the body. In order to achieve swimming in 3D space, the researchers proposed pectoral fins have a motion with 2 DOF. The thrust forces of the pectoral fins have been evaluated by several experiments. The proposed design has been simulated by the flapping motion of the fin as two waves, and then the proposed design was practically validated.

In 2017, Behbahani and Tan in Michigan State University, USA proposed the dynamical model of a robotic fish, the design was achieved based on Labriform swimming mode. The swimming of this robot was achieved by two flat pectoral fins, which generated forward thrust force. The performance of the robot fins was improved by adding flexibility for these fins. The performance of the flexible pectoral fins was tested for different values of flexibility. Further, this robot had a caudal fin but the effectiveness of this fin was neglected which was used for balancing only.

In 2018, Chinese researchers proposed a swimming robot, which can achieve a good performance steering process based on 2 DOF pectoral fins and flexible caudal fin. In this study, the hydrodynamics of the swimming robot was evaluated first time for three cases of thrust forces by caudal fin, pectoral fins, and both caudal and pectoral fins. The steering characteristics of these three cases have been simulated, and the results of this study show that the thrust force is optimized by cooperation between caudal and pectoral fins in propelled process.

Fish during swimming can increase their forward speed by tuning the thrust/drag forces of the pectoral fins, which leads to an increased difference between power stroke and recovery stroke in rowing mode. In 2019, researchers from Vietnam and Korea have proposed a swimming robot that has folding pectoral fins with one degree of freedom, which is employed to generate forward thrust. Each pectoral fin is connected to the robot body through a flexible joint, which increases the power stroke with respect to recovery stroke. The Morison force model has been used to represent the dynamic interaction between pectoral fin and fluid. For this study, several experiments have been done in order to validate the proposed design, while a practical experimental platform is implemented for practical validation. The results show the success of the proposed design by evaluating the force of the forward thrust.

The swimming performance of organisms underwater is dependent on the biomechanical structure and the shape of the fins. Therefore in 2020, a group of researchers from the University of Basrah, Iraq, have proposed a new design of swimming robot based on the observation of Labriform mode swimming organism, where the pectoral fins of this robot are designed as a concave shape in order to increase the ratio between the power stroke and the recovery stroke. Three pectoral fin types of different sizes are implemented to select the optimum design that generates maximum thrust force. A maximum thrust force is achieved at the maximum difference between power strokes to recovery stroke. The hydrodynamic model of the proposed design is derived and analyzed by computational fluid dynamics (CFD). The simulation of the proposed design is achieved by the SOLIDWORKS® platform, and then practically validated by implementing several experiments. The results of the simulation and practical experiments show the success of the proposed design of the swimming robot.

1.3. Organization of Chapters

After the introductory chapter that includes concepts of Labriform swimming robot, the objective of this book is to include a literature survey about Labriform swimming robot, the chapters of the book will be as follows:

1.3.1. *Chapter 2*

To design a pectoral fin-based robot, the main locomotion of the robot will depend on the parameter known as the aspect ratio (Ar), which represents the ratio between the squared length of the fin from the base to the tip, to the projected surface area facing the water. By observation, the experimental results of the research show that the Labriform-based fish with a low Ar of paddle-shaped swims slowly and occupy fewer energetic areas in contrast to the Labriform-based with high Ar of wing-shaped fins. Therefore, the work provided within this chapter represents the first step in designing a Labriform swimming robot with pectoral fins only. The effect of concave pectoral fins in producing the highest thrust while keeping the drag force to a minimum has been studied carefully. With the aid of Computational Fluid Dynamics (CFD) analysis, the optimum shape and size from three distinct concave pectoral fins have been discussed, each one with a specific aspect ratio Ar.

One of the main issues in producing forward propulsion is the difference between thrust force and drag force. Therefore, the design of pectoral fins as concave assists in maximizing the thrust force to a maximum, when the fin pushes backward toward the rear part of the body, minimizing the drag force to a minimum when the fin returns, which produces the maximum difference between thrust force and drag force.

To swim at a constant speed, the total hydrodynamics (including drag/lift forces and momentums) acting on a fish should be balanced according to the principle of momentum conservation. Different parameters determine the momentum exchange like Reynolds number, reduced frequency, and overall geometrical shape. These parameters will be evaluated and studied to optimize the fin dimensions.

To validate using the concave shape of pectoral fins for Labriform swimming robot, several experiments were implemented to evaluate the performance of these fins starting from simulation by SOLID-WORKS® platform and MATLAB to practical implementation by design prototype for these fins.

1.3.2. *Chapter 3*

As discussed in Chapter 2, the issues of pectoral fin shape, size, and biomechanics have a significant impact on the swimming performance of the Labriform swimming robot. In addition, the shape, size, and body weight of a swimming robot affect the swimming motion. Recently, one of the main areas of interest to researchers in robotics is how to solve some of the difficulties associated with different aquatic environments.

Therefore, this chapter includes the design and implementation of a swimming robot propelled by two pectoral fins, in which the full-body design of the swimming robot is presented. Then, the effect of the changing fin oscillation speed will be observed in two stages, the first stage is the first force stroke, in which the fins begin to push back from the body, and the recovery stage, in which the fins return towards the forward part of the body. The detailed steps of robot design are shown and the thrust exerted by the pectoral fins will be evaluated. The proposed swimming robot is a rigid body with an oval cross-sectional area, which helps reduce water resistance during the propulsion process.

For the robust design of the swimming robot, the kinematic and dynamic models must be derived, as achieved by Newton-Euler equations, moreover, the hydrodynamic forces exerted on the swimming robot are evaluated using the method of CFD. A PID controller will be used to improve the movement performance of the swimming robot. The swimming robot design will be validated by simulating the swimming robot design in the SOLIDWORKS® platform and MATLAB, while several hands-on practical experiments will be conducted for physical testing.

1.3.3. *Chapter 4*

Directional control is related to maneuverability. Maneuverability is an important aspect of the locomotor performance of aquatic animals and underwater vehicles, such as swimming robots. Maneuvering is an integral part of steering, course correction, obstacle avoidance in a complex environment, stabilizing in a high-energy ocean, jumping and diving, tracking targets, and swimming through active turbulent waters. Maneuver studies have focused on lateral steering (i.e. yawning). Aquatic animals show high performance in maneuverability while sacrificing little stability. For example, a fish can achieve good speed while steering in a radius of 10–30% of its body length.

Narrow and fast turning are distinguished by maneuverability and agility, respectively. Maneuverability is the ability to turn in a narrow space and is measured as the ratio of steering radius to the length of the body (R/BL, where R is the turning radius and BL is the total length of the body). The objective of this chapter is to present two studies on the routing strategies used in developing the Labriform Mechanism. These two strategies include studying the ability of the rigid body of boxfish to achieve high performance in the guiding process based on the caudal fin, while the other study includes modeling and experimental validation of this model to achieve the steering process by the pectoral fins. While the thrust force in these two systems is generated by pectoral fins.

For the first system, although the boxfish has rigid armor that limits body undulation, it is highly maneuverable and can swim with remarkable dynamic stability. The turning torque in the flow basin will be measured using a physical model with attachable caudal fin closed or open at the different body and tail angles, and different water flow velocities, as the results are evaluated by the simulated CFD, which indicates that the caudal fin is achieve steering. The caudal fin acts as a rudder for a rigid body that is naturally unstable for steering. Boxfishes appear to use unstable body interaction and active changes in the shape and direction of their caudal fin to modify their maneuverability and stability. While, the second system aims to develop a swimming robot with good steering performance, in which the steering behavior is achieved by one degree of freedom (1-DOF) represented by two concave-shaped pectoral fins. The steering mechanism adopted here is based on the differential drive principle. This principle is carried out by varying the right/left fin velocities. Different radii achieve four different cases of velocities. The proposed design will be validated theoretically via the SOLIDWORKS® platform and then will be proved practically in a physical swimming pool.

1.3.4. *Chapter 5*

During the diving process, the thrust must be up and down to move the swimming robot up and down in the water. This process can be achieved by controlling the buoyancy of the swimming robot using an autonomous mechanism. Therefore, the vertical movement will be generated by changing the buoyancy of the swimming robot, which converts the 2 DOF movement of the swimming robot into a 3 DOF movement.

In general, swimming robots have higher maneuverability, for example, with respect to swimming turning radius than autonomous underwater vehicles (AUV), but they require constant swimming operation, and they cannot operate for a long time compared to AUVs. The robot diving concept has been developed by combining the desirable characteristics of an AUV with a swimming robot in one design. However, integrating the fin operating mechanisms, with the principle of controlling the center of gravity (CoG) is key to overcome the diving challenge.

This chapter presents the design and modeling of a swimming robot that can realize the diving process based on the sliding block mechanism, while the diving systems that will be presented in this chapter overcome the challenges of scale in swimming robots, two systems that overcome the challenges of scale. Propulsion of the first swimming robot is based on the Carangiform mechanism in swimming, the second is based on the principle of the Labriform mechanism, while the diving system in these two models is based on the Control Center of Gravity (CoG) using a sliding block system.

Several experiments will be introduced to validate these systems, in which SOLIDWORKS® platform and MATLAB will be employed for simulation experiments and then practical experiments will be designed in order to compare the practical and theoretical results.

1.3.5. *Chapter 6*

This chapter includes the conclusions of the book from stages of designing a swimming robot to identify the strengths and drawbacks of the systems used in manufacturing such robots.

References

Ali, A. A., Fortuna, L., Frasca, M., Rashid, M. T. and Xibilia, M. G. (2011). Complexity in a population of artemia, *Elsevier Journal, Chaos, Solitons & Fractals*, 44(4-5), pp. 306–316.

Anderson, J. M., Streitlien, K., Barrett, D. S. and Triantafyllou, M. S. (1998). Oscillating foils of high propulsive efficiency, *Journal of Fluid Mechanics*, 360, pp. 41–72.

Bandyopadhyay, P. R., Beal, D. N. and Menozzi, A. (2008). Biorobotic insights into how animals swim, *Journal of Experimental Biology*, 211(2), pp. 206–214.

Bazaz Behbahani, S., Wang, J. and Tan, X. (2013). A dynamic model for robotic fish with flexible pectoral fins, *In IEEE/ASME International Conference on Advanced Intelligent Mechatronics (AIM)*, pp. 1552–1557, (Wollongong, Australia).

Behbahani, S. B. and Ta, X. (2016). Bio-inspired flexible joints with passive feathering for robotic fish pectoral fins, *Bioinspiration & Biomimetics*, 11(3).

Behbahani, S. B. and Tan, X. (2017). Role of pectoral fin flexibility in robotic fish performance, *J. Nonlinear Sci.*, 27(4), pp. 1155–1181.

Breder, C. M. (1924). Respiration of factor in locomotion of fishes, *The American Naturalist*, 58(655), pp. 145–155.

Castaño, M. L. and Tan, X. (2019). Model predictive control-based path-following for tail-actuated robotic fish, *ASME. J. Dyn. Sys., Meas., Control*, 141(7).

Chiu, F. C., Chen, C. K. and Guo, J. (2004). A practical method for simulating pectoral fin locomotion of a biomimetic autonomous underwater vehicle, *In International Symposium on Underwater Technology (UT)*, pp. 323–329, (Taipei, Taiwan).

Coral, W., Rossi, C., Curet, O. M. and Castro, D. (2018). Design and assessment of a flexible fish robot actuated by shape memory alloys, *Bioinspiration & Biomimetics*, 13(5).

Deng, X. and Avadhanula, S. (2005). Biomimetic micro underwater vehicle with oscillating fin propulsion: System design and force measurement, *In IEEE International Conference on Robotics and Automation (ICRA)*, pp. 3312–3317, (Barcelona, Spain).

Dong, H., Bozkurttas, M., Mittal, R., Madden, P. and Lauder, G. V. (2010). Computational modelling and analysis of the hydrodynamics of a highly deformable fish pectoral fin, *Journal of Fluid Mechanics*, 645, pp. 345–373.

Fortuna, L., Frasca, M., Xibilia, M. G., Ali, A. A. and Rashid, M. T. (2010). Motion Control of A Population Of Artemias, *IEEEXplore, The 1st International Conference on Energy, Power, and Control, Basrah, Iraq*, pp. 12–15.

Kanso, E., Marsden, J. E., Rowley, C. W. and Melli-Huber, J. B. (2005). Locomotion of articulated bodies in a perfect fluid, *Journal of Nonlinear Science*, 15(4), pp. 255–289.

Kato, N. and Furushima, M. (1996). Pectoral fin model for maneuver of underwater vehicles, *In Symposium on Autonomous Underwater Vehicle Technology (AUV)*, pp. 49–56, (Monterey, CA, USA).

Kato, N. and Inaba, T. (1998). Guidance and control of fish robot with apparatus of pectoral fin motion, *In IEEE International Conference on Robotics and Automation (ICRA)*, 1, pp. 446–451, (Leuven, Belgium).

Kelly, S. D. and Murray, R. M. (2000). Modelling efficient pisciform swimming for control, *International Journal of Robust and Nonlinear Control*, 10(4), pp. 217–241.

Kim, B., Kim, D. H., Jung, J. and Park, J. O. (2005). A biomimetic undulatory tadpole robot using ionic polymermetal composite actuators, *Smart Materials and Structures*, 14(6), pp. 1579.

Kopman, V. and Porfiri, M. (2013). Design, modeling, and characterization of a miniature robotic fish for research and education in biomimetics and bioinspiration, *IEEE/ASME Transactions on Mechatronics*, 18(2), pp. 471–483.

Lachat, D., Crespi, A. and Ijspeert, A. J. (2006). BoxyBot: A swimming and crawling fish robot controlled by a central pattern generator, *In IEEE/RAS-EMBS International Conference on Biomedical Robotics and Biomechatronics (BIOROB)*, pp. 643–648, (Pisa, Italy).

Lauder, G. V. and Tangorra, J. L. (2015). Fish locomotion: Biology and robotics of body and fin based movements, *In Robot Fish, ser. Springer Tracts in Mechanical Engineering*, pp. 25–49.

Li, G., Deng, Y., Osen, O. L., Bi, S. and Zhang, H. (2016). A bio-inspired swimming robot for marine aquaculture applications: From concept-design to simulation, *OCEANS*, pp. 1–7.

Li, Z., Ge, L., Xu, W. and Du, Y. (2018). Turning characteristics of biomimetic robotic fish driven by two degrees of freedom of pectoral fins and flexible body/caudal fin, *International Journal of Advanced Robot Systems*, 15(1).

Liu, H., Wassersug, R. and Kawachi, K. (1996). A computational fluid dynamics study of tadpole swimming, *Journal of Experimental Biology*, 199(6), pp. 1245–1260.

Liu, J. and Hu, H. (2010). Biological inspiration: From carangiform fish to multi-joint robotic fish, *Journal of Bionic Engineering*, 7(1), pp. 35–48.

Ma, H., Cai, Y., Wang, Y. *et al.* (2015). A biomimetic cownose ray robot fish with oscillating and chordwise twisting flexible pectoral fins, *Industrial Robot: An International Journal*, 42(3).

Marchese, A. D., Onal, C. D. and Rus, D. (2014). Autonomous soft robotic fish capable of escape maneuvers using fluidic elastomer actuators, *Soft Robot*, 1(1).

Meurer, C., Simha, A., Kotta, U. and Kruusmaa, M. (2019). Nonlinear orientation controller for a compliant robotic fish based on asymmetric actuation, *International Conference on Robotics and Automation (ICRA)*, pp. 4688–4694, (Montreal, QC, Canada).

Mihalitsis, M. and Bellwood, D. (2019). Morphological and functional diversity of piscivorous fishes on coral reefs, *Coral Reefs*, 38, pp. 945–954.

Ming, W., Zhi, Y. J., Min, T. and Wei, Z. J. (2012). Multimodal swimming control of a robotic fish with pectoral fins using a CPG network, *Mechanical Engineering*, 57(10), pp. 1209–1216.

Morgansen, K. A., Fond, T. M. L. and Zhang, J. X. (2006). Agile maneuvering for fin-actuated underwater vehicles, *In Second International Symposium on Communications, Control and Signal Processing*, (Marrakech, Morocco).

Morgansen, K. A., Triplett, B. I. and Klein, D. J. (2007). Geometric methods for modeling and control of free-swimming fin-actuated underwater vehicles, *IEEE Transactions on Robotics*, 23(6), pp. 1184–1199.

Naser, F. A. and Rashid, M. T. (2019). Design, modeling, and experimental validation of a concave-shape pectoral fin of labriform-mode swimming robot, *Engineering Reports*, 1(5), pp. 1–17.

Naser, F. A. and Rashid, M. T. (2020). Effect of Reynold number and angle of attack on the hydrodynamic forces generated from a bionic concave pectoral fins, *IOP Conf. Ser.: Mater. Sci. Eng.*, 745, pp. 1–13.

Naser, F. A. and Rashid, M. T. (2020). The influence of concave pectoral fin morphology in the performance of labriform swimming robot, *Iraqi Journal for Electrical and Electronic Engineering*, 16(1), pp. 54–61.

Naser, F. A. and Rashid, M. T. (2020). The influence of concave pectoral fin morphology in the performance of labriform swimming robot, *Iraqi Journal for Electrical and Electronic Engineering*, 16(1).

Naser, F. A. and Rashid, M. T. (2021). Design and realization of labriform mode swimming robot based on concave pectoral fins, *Journal of Applied Nonlinear Dynamics*, 10(4), pp. 691–710.

Naser, F. A. and Rashid, M. T. (2021). Enhancement of labriform swimming robot performance based on morphological properties of pectoral fins, *J. Control. Autom. Electr. Syst.*, 32, pp. 927–941.

Naser, F. A. and Rashid, M. T. (2021). Implementation of steering process for labriform swimming robot based on differential drive principle, *Journal of Applied Nonlinear Dynamics*, 10(4), pp. 737–753.

Naser, F. A. and Rashid, M. T. (2021). Labriform swimming robot with steering and diving capabilities, *Journal of Intelligent & Robotic Systems*, 103(14), pp. 1–19.

Newman, J. N. and Wu, T. Y. (1975). Hydromechanical aspects of fish swimming, *In Swimming and Flying in Nature, Springer US*, pp. 615–634.

Pfeil, S., Katzer, K., Kanan, A., Mersch, J., Zimmermann, M., Kaliske, M. and Gerlach, G. (2020). A biomimetic fish fin-like robot based on textile reinforced silicone, *Micromachines (Basel)*, 11(3), pp. 298.

Pham, V., Nguyen, T., Lee, B. and Vo, T. (2020). Dynamic analysis of a robotic fish propelled by flexible folding pectoral fins, *Robotica*, 38(4), pp. 699–718.

Phelan, C., Tangorra, J. and Lauder, G. (2010). A biorobotic model of the sunfish pectoral fin for investigations of fin sensorimotor control, *Bioinspiration & Biomimetics*, 5(3).

Popovic, M. B. (2013). Biomechanics and robotics, (CRC Press).

Raj, A. and Thakur, A. (2016). Fish-inspired robots: Design, sensing, actuation, and autonomy — a review of research, *Bioinspiration & Biomimetics*, April 19.

Rashid, M. T., Ali, A. A., Ali, R. S., Fortuna, L., Frasca, M. and Xibilia, M. G. (2012). Wireless Underwater Mobile Robot System Based on ZigBee, *IEEEXplore, 2012 International Conference on Future Communication Networks, Baghdad, Iraq*, pp. 117–122.

Rashid, M. T., Frasca, M., Ali, A. A., Ali, R. S., Fortuna, L. and Xibilia, M. G. (2012). Artemia swarm dynamics and path tracking, *Nonlinear Dynamic*, 68, pp. 555–563.

Rashid, M. T., Frasca, M., Ali, A. A., Ali, R. S., Fortuna, L. and Xibilia, M. G. (2012). Nonlinear model identification for artemia population motion, *Nonlinear Dynamic*, 69, pp. 2237–2243.

Rashid, M. T. and Rashid, A. T. (2016). Design and implementation of swimming robot based on labriform model, *Al-Sadeq International Conference on Multidisciplinary in IT and Communication Science and Applications (AIC-MITCSA)*, pp. 1–6.

Rashid, M. T., Naser, F. A. and Mjily, A. H. (2020). Autonomous micro-robot like sperm based on piezoelectric actuator, *International Conference on Electrical, Communication, and Computer Engineering (ICECCE)*, pp. 1–6.

Ren, Q., Xu, J., Yang, S. and Yan, L. (2014). Design and implementation of a biomimetic robotic fish with 3D locomotion, *11th IEEE International Conference on Control & Automation (ICCA)*, (Taiwan).

Saab, W., Rone, W. and Ben-Tzvi, P. (2018). Discrete modular serpentine robotic tail: Design, analysis and experimentation, *Robotica*, 36(7), pp. 994–1018.

Salazar, R., Campos, A., Fuentes, V. and Abdelkefi, A. (2019). A review on the modeling, materials, and actuators of aquatic unmanned vehicles, *Ocean Engineering*, 172, pp. 257–285.

Sfakiotakis, M., Lane, D. M. and Davies, J. B. C. (1999). Review of fish swimming modes for aquatic locomotion, *IEEE Journal of Oceanic Engineering*, 24(2), pp. 237–252.

Sitorus, P. E., Nazaruddin, Y., Leksono, E. and Budiyono, A., (2009). Design and implementation of paired pectoral fins locomotion of labriform fish applied to a fish robot, *Science Direct, Journal of Bionic Engineering*, 6, pp. 37–45.

Su, Z., Yu, J., Tan, M. and Zhang, J. (2014). Implementing flexible and fast turning maneuvers of a multijoint robotic fish, *IEEE/ASME Transactions on Mechatronics*, 19(1).

Tan, X. (2011). Autonomous robotic fish as mobile sensor platforms: Challenges and potential solutions, *Marine Technology Society Journal*, 45(4), pp. 31–40.

Tan, X., Kim, D., Usher, N., Laboy, D., Jackson, J., Kapetanovic, A., Rapai, J., Sabadus, B. and Zhou, X. (2006). An autonomous robotic fish for mobile sensing, *In IEEE/RSJ International Conference on Intelligent Robots and Systems (IROS)*, pp. 5424– 5429.

Tehaní, C., Naula, E. A., Garza-Castañón, L. E., Vargas-Martínez, A., Martínez-López, J. I. and Minchala-Ávila, L. I. (2020). Design, construction, and modeling of a BAUV with propulsion system based on a parallel mechanism for the caudal fin, *Appl. Sci.*, 10(7), pp. 2426.

Triantafyllou, M. S., Triantafyllou, G. S. and Yue, D. K. P. (2000). Hydrodynamics of fishlike swimming, *Annual Review of Fluid Mechanics*, 32(1), pp. 33–53.

Wang, J. and Tan, X. (2013). A dynamic model for tail-actuated robotic fish with drag coefficient adaptation, *Mechatronics*, 23(6), pp. 659–668.

Wang, W. and Xie, G. (2014). Cpg-based locomotion controller design for a box fish-like robot, *International Journal of Advanced Robotic Systems*, 11(6).

Watanabe, T., Watanabe, K. and Nagai, I. (2018). Thrust analysis of propulsion mechanisms with pectoral fins in a Manta robot, *IOP Conference Series: Materials Science and Engineering*, 619, (South Sulawesi, Indonesia).

Zhang, F., Ennasr, O., Litchman, E. and Tan, X. (2016). Autonomous sampling of water columns using gliding robotic fish: Algorithms and harmful-algae-sampling experiments, *IEEE Systems Journal*, 10(3), pp. 1271–1281.

Zhang, S., Qian, Y., Liao, P., Qin, F. and Yang, J. (2016). Design and control of an agile robotic fish with integrative biomimetic mechanisms, *IEEE/ASME Transactions on Mechatronics*, 21(4), pp. 1681–1688.

Zhang, Y. H., He, J. H., Yang, J., Zhang, S. W. and Low, K. H. (2006). A computational fluid dynamics (CFD) analysis of an undulatory mechanical fin driven by shape memory alloy, *International Journal of Automation and Computing*, 3(4), pp. 374–381.

Zhao, W., Ming, A. and Shimojo, M. (2018). Development of high-performance soft robotic fish by numerical coupling analysis, *Appl. Bionics Biomech.*, 2018, pp. 1–12.

Zhong, Y., Li, Z. and Du, R. (2018). Robot fish with two-DOF pectoral fins and a wire-driven caudal fin, *Advanced Robotics*, 32.

Zhu, J., White, C., Wainwright, D. K., Di Santo, V., Lauder, G. V., Bart-Smith, H. (2019). Tuna robotics: A high-frequency experimental platform exploring the performance space of swimming fishes, *Science Robotics*, 4(34).

Chapter 2

DESIGN AND VALIDATION OF THE PECTORAL FINS

2.1. Introduction

The study of real fish by observing their motion during swimming inside water is a good source of modeling the swimming motion of these fish, and the fish can be classified according to swimming modes. The modes of swimming depend on the types of fins of fish, further to employing methods of these fins during swimming. The fins of fish are different concerning their size, shape, location, and flexibility. Hence, there are two types of swimming of fish according to the use of fins. As was mentioned in the previous chapter, these fins use the body and caudal fin (BCF) or the median and paired fins (MPF).

Since the fish is classified with respect to the fin types, therefore, the motion of the fins can be further classified into the undulatory type and oscillatory type. The undulatory type generates a wave-like motion on a fin surface while the oscillatory type generates an oscillating motion. Based on the classification of the motions fin, the swimming mechanisms can be categorized into different swimming modes. Generally, fish may use more than one swimming mechanism, either simultaneously or at particular speeds.

2.2. The Description of Fins

Fish use different kinds of fins for achieving the locomotion process. These types of fins are named Dorsal fin, Pelvic fin, Caudal fin,

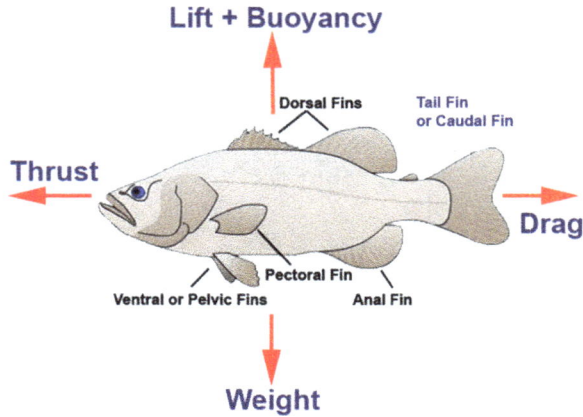

Figure 2.1. Types of fish fins.

Anal fin, and Paired pectoral fins as shown in Figure 2.1. It is impossible to find fish that uses all these fins at the same time. Only some are used as the main locomotors or a combination of more than one fin may be used. The function of each fin depends on the motion of the fish and its environment.

Either dorsal fins are located on the lower back of the fish or at the top; they assist the fish during sharp turnings or sudden stops. Fish may also have up to three different kinds of Dorsal fins, usually known as proximal, middle, and distal fins, on the other hand, many fish have just a couple of Dorsal fins with the middle and distal fins combined. Dorsal fin types are Single, Split, Pointed, Trigger, Spine Triangular, and Trailing.

A couple of Ventral or Pelvic fins are located on the bottom front of the fish, which may help in fish stability and are useful during sudden stopping and slowing down of the fish velocity. The Caudal fin or generally known as a Tail fin can be considered as the main fin that generates the forward thrust and increasing up-velocity in many types of fish. In bony fish, there are many different sorts of Caudal fins known as Indented, Round, Square, Forked, Lunate, and Pointed. The Anal fin helps the Dorsal fins by making them more stable in the water. Pectoral fin shape has important influence on swimming hydrodynamics, usual habitat uses, and energetics as mentioned.

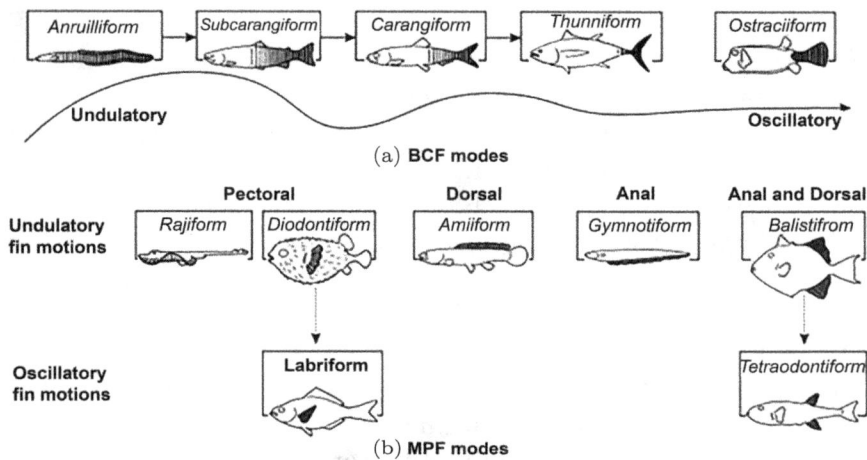

Figure 2.2. Swimming modes (a) BCF. (b) MPF [19].

The categories of swimming mechanism modes are described in Figure 2.2, which shows that different fish use different swimming modes, for BCF they are Anguilliform, Subcarangiform, Carangiform, Thunniform, and Ostraciiform. The motion of these modes ranges from undulatory to oscillatory motion as shown in Figure 2.2(a). Other modes like Rajiform, Diodontiform, Amiiform, Gymnotiform, and Balastiform are MPF that generated motion of undulatory type of MPF, whereas Labriform and Tetraodontifom are of MPF of oscillating type as can be seen in Figure 2.2(b).

The description of the modes of swimming mechanisms in the fish will be summarized as follows:

2.2.1. *Propulsion by Undulation of Median or Pectoral Fins*

Both forward and backward propulsions usually are generated by pectoral fins undulation; furthermore, reverse the direction, and the ability of maneuverable. In addition, the body axis will remain straight, this is unavoidable for fish whose bodies are neither bendable nor flexible, and it may require certain electrosensory or acoustico-lateralis systems operation.

2.2.1.1. *Amiiform Mode*

In 1926, Breder used the name Amiiform for the fish swimming by undulation of the long dorsal fin, for example, Amia calva as shown in Figure 2.3. This fish is covered by heavy shield of shells, usually, this fish uses its dorsal fin for locomotion, although, the locomotion is supported by undulating its body based on Subcarangiform mode. It has a complete set of developed paired and median fins.

2.2.1.2. *Gymnotiform Mode*

The swimming of Gymnotus carupo fish is an example of the gymnotiform mode, in which there is no dorsal fin in this type of fish compared with Gymnarchus swimming in the amiiform mode (see Figure 2.4). The propulsion force of forward and backward motion is generated by rapid undulate along the elongate anal fin, which has a short wavelength in either direction. During this motion the body can hold straight. Further, the fish can support this thrust force by undulating the body with an oscillation frequency about four times the oscillation frequency generated by the fin (Hertel, 1966).

2.2.1.3. *Balastiform Mode*

The swimming by the undulation of anal and dorsal fins is known as balastiform. Examples of fish that belong to the family Balisdae are filefish and triggerfish (see Figure 2.5), in which the oscillation force of their fins generates forward thrust. These fish can reverse the

Figure 2.3. Amia calva fish.

Figure 2.4. Gymnotus carapo fish.

Figure 2.5. Balisdae fish.

direction of motion by reversing the direction of oscillation of their fins. Further, the diving process is achieved by passing the oscillation wave backward on the anal fin and forward on the dorsal fin, while the buoyancy process is achieved by reversing these oscillation wave directions. The propulsion force can also be increased to maximum speed by changing the caudal fin from wrapped to extended with tail of 2.5 times.

2.2.1.4. *Rajiform Mode*

Most fish of the ray family employ Rajiforms mode in their primary locomotor, in which their pectoral fins are enlarged in a wide lateral expansion of the body. Undulating of the fin generates waves in the vertical plane, which pass backward in parallel with pectoral fin margins as shown in Figure 2.6. In the rays-type fish, only the pelvic fins contribute with the pectoral fins in the forward locomotion process. The pelvic fins are located at the posterior of the disc created by the pectoral fins and snout, these fins are further used for locomotion backward by kicking back from the bottom.

The mantas, Mobulidae, and Myliobatidae belong to the eagle rays families that have pectoral fins. These fins expand more in Rajidae, therefore their width is greater than the length of the body. The locomotion of these organisms underwater produces a high-performance rajiform swimming motion that seems similar to the birds of flight.

2.2.1.5. *Diodontiform Mode*

The principle of motion of porcupine-fishes that belong to the diodontiform is achieved by the anguilliform undulation produced by

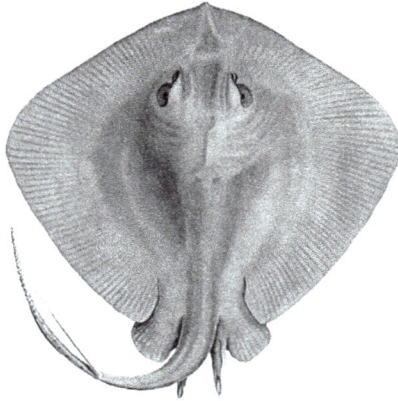

Figure 2.6. Rays fish.

Figure 2.7. Porcupine-fishes.

the moderately broad pectoral fins. The pectoral fins of diodontiform swimmers are localized in a variable plane that may be near vertical as shown in Figure 2.7. The oscillation waves produced by the undulatory of pectoral rays on their bases may have a vertical component, partly ignored based on the fanlike divergence of the rays. Besides, the opposite pectoral fins diverge from the body laterally at the complementary angles, which can be changed. Thrust force in any direction may be generated by combining with the flapping of the whole fin in the labriform mode; this will lead to very slow motion but with precise maneuvering.

2.2.2. *Propulsion by Oscillation of Median or Pectoral Fins*

These are very short-based fins, where their oscillation can be compared with ostraciiform propulsion. Several intermediate fin modes can also be compared with carangiform or other modes, depending on the fin length and the number of wavelengths. This comparison is not sufficient, due to short fins being capable of rotating on the base to generating locomotive strokes without the body/caudal fin series of modes.

2.2.2.1. *Tetraodontiform Mode*

In the family Tetraodontidae, puffer fish use their anal and dorsal fins, as a unit, similar to paddle in flapping to produce forward propulsion as shown in Figure 2.8. Usually the caudal fin is used for steering, while the depressor and erector muscles of the anal and dorsal expand and contract to move the fin rays laterally.

2.2.2.2. *Labriform Mode*

Some fish use the pectoral fins in swimming, which are fan-like and of circular shape. These fish swim based on the mode named Labriform (see Figure 2.9), where the pectoral fins flap to produce forward propulsion. In the Mudminnow Umbra and the Stickleback Gasterosteus, the pectoral fins may flap quickly, which generates high frequency oscillation, in this case, the fins appear as a blur. During the flapping of pectoral fins, there are two forces produced as

Figure 2.8. Puffers.

Figure 2.9. Labrus merula fish.

power stroke and recovery stroke. These forces are the thrust force and drag force, which produce forward displacement and backward displacement, while the overall displacement of swimming is the difference between these two values.

2.3. Employing Pectoral Fin as Locomotors

The oscillating movements of pectoral fins produce thrust force in the Labriform swimming mechanism, where there are three types of oscillation in this mode as shown in Figure 2.10, as:

(1) Rowing motion, which is considered a drag-based motion.
(2) Flapping motion, considered as lift-based motion, which seems similar to flying of bird.
(3) Feathering motion, which is a combination of rowing and flapping motions.

Usually, at small value speeds, the drag-based motion is more efficient when the flow of chordwise fin is small. While, at larger speeds, they are more likely to be efficient in lift-based movement. Pectoral fins provide a significant function in the swimming maneuvering capability and stability of real fish when generating or helping in the propulsion mechanism.

To design a pectoral fin-based robot, the main locomotion of the robot will depend on the parameter known as the aspect ratio (Ar), which represents the ratio between the squared length of the fin from base to tip, to the projected surface area facing the water. Observation and experimental results of research show that the Labriform-based fish with a low Ar of paddle-shape, swims slowly

Figure 2.10. Three types of pectoral fin motions.

and occupy fewer energetic areas in contrast to the Labriform-based with high Ar of wing-shaped fins. Practical studies have affirmed emphatically that Labriform-based with wing-like design fins use a flapping swimming mode while Labriform-based with paddle-like design fins possibly use rowing or combined swimming mode. A combination of these two studies of flapping and rowing contributes to the rowing-flapping model study.

It is worthy of realizing that fish that swims with pectoral fin flapping can achieve higher speeds than BCF swimmers of equal size, wherein the rowing pectoral fins have lower performance than expected. Labriform mode with pectoral fin movements with steady swimming is categorized as Hydrolagus, Gomphosus, Cirrhilabrus, Tautoga, Scarus, and Lactoria, Tetrasomus.

Water density and incompressibility are the main characteristics of a locomotion environment; these two properties play a vital role in fish evolution. The water is considered as an incompressible fluid, therefore, a motion that occurs due to an underwater creature will let the surrounding water move with it. The density of the water overrides that of air by 800 times. Swimming includes momentum exchange involving the fish and the water that surrounds it. The essential momentum exchange occurs through drag, lift, and reaction forces. The drag force is generated by the following components:

(1) Skin friction is known as viscous (friction drag), which occurs between the surrounding boundary layer of water and the fish.
(2) Pressures are known as form drag, which is generated when pushing water away by the fish to pass.

(3) Induced drag force, which is known as vortex drag. This drag occurs when the energy is spent in the formed vortices generated by the tail and pectoral fins. The last two components are combined together in the description as pressure drag.

2.4. Optimum Design of Pectoral Fins

One of the main issues in producing forward propulsion is the difference between thrust force and drag force. Therefore, the design of pectoral fins as concave assists in maximizing the thrust force to a maximum, when the fin pushes backward toward the rear part of the body, minimizing the drag force to a minimum when the fin returns, which produces the maximum difference between thrust force and drag force.

In order to swim at a constant speed, the total hydrodynamics (including drag/lift forces and momentums) acting on a fish should be balanced according to the principle of momentum conservation as shown in Figure 2.1. Different parameters determine the momentum exchange like Reynolds number, reduced frequency, and overall geometrical shape. The value of Reynolds number (Re) can be considered as the ratio of the inertial forces to the viscous forces as given below:

$$\text{Re} = \frac{LU}{\lambda} \tag{2.1}$$

where L is the length of the swimming fish, U is the swimming velocity of the fish and λ is water kinematic viscosity. Generally, for Re in an adult fish, where inertial forces prevail over the viscous forces that are neglected, the effective forces can be generated by the incorporation of Re pressure, acceleration reaction, drag force, and lift force mechanisms.

On the other hand, the reduced frequency σ refers to the significance of time-dependent state influences on the water flow, which is expressed as follows:

$$\sigma = 2\pi \frac{fL}{U} \tag{2.2}$$

where f is a notation to the oscillation frequency. The reduced frequency analyzes the time consumed by a particle to overcome the length of an object to the time required to finish one cycle of movement. Besides, it is considered as a measure of acceleration response (reaction) to pressure drag force and lift force. When $\sigma < 0.1$, here, it is a steady motion and acceleration reaction forces have a small impact. But if $0.1 < \sigma < 0.4$, in this case, all force generation mechanisms will be essential, furthermore at larger values of σ, acceleration reaction will dominate. Experimentally, the reduced frequency is seldom found below the threshold of 0.1.

The other parameter that determines the momentum exchange is the shape of the swimming fish, which largely affects the magnitude of the force components. There exist great number of studies on the relationship in steady state lift and drag forces generation, on the other hand, few works have been done on the association between shape and acceleration response. One on swimming efficiency measurement is Froude efficiency where the efficiency $\hat{\eta}$ is given as:

$$\hat{\eta} = \frac{\langle T \rangle U}{\langle P \rangle} \tag{2.3}$$

U is the forward velocity of the fish, $\langle T \rangle$ is the averaged generated thrust, and $\langle P \rangle$ is the averaged power required over time.

On the other hand, another efficiency measure of the pectoral fin shape effect is the Aspect ratio (Ar) (see Figure 2.11). It can be expressed by dividing the squared value of the pectoral fin length, by the projected surface area that is facing water as follows:

$$Ar = \frac{PF_L^2}{S} \tag{2.4}$$

where PF_L is the pectoral fin length and S is the projected surface area. By giving an estimate of the turning of fin shape, these measurements are directly connected to forward velocity performance and thrust generation.

Tapered, high Ar fins are generally found in fish to generate thrust due to the lift force, keeping high velocities, whereas rounded, with low Ar fins, are related to the rowing motion fin, generating

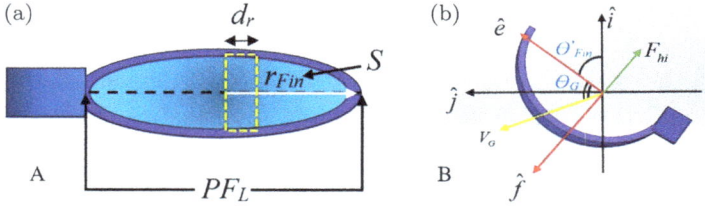

Figure 2.11. Proposed pectoral fin design. (a) Front view. (b) Top view.

thrust due to the drag force and are usually used for low speed maneuverability.

Another factor that benefits swimming efficiency is Strouhal number (St), for biological fish, generally lying in the range of $(0.05–0.6)$. It is a dimensionless number that describes the oscillating flow mechanism. St is defined as:

$$St = \frac{2f(PF_L)\sin A}{|V_b|} \qquad (2.5)$$

where A is the fin oscillating amplitude from peak to peak and V_b is the forward body velocity. The hydrodynamic forces F_{hi} (representing F_D, F_L and M_D, that are described in the next section) are due to the proposed pectoral fins as shown in Figure 2.11(b). These forces can be defined as:

$$\left.\begin{array}{l} F_D = 1/2\rho|V_b|^2\,SC_D \\ F_L = 1/2\rho|V_b|^2\,SC_L \\ M_D = -C_M w_{\mathrm{Fin}}^2\mathrm{sgn}(w_{\mathrm{Fin}}) \end{array}\right\} \qquad (2.6)$$

where F_D, F_L and M_D are the drag force, lift force and momentum force respectively. ρ is the water density, $|V_b|$ indicates the magnitude value of the body's linear velocity where $|V_b| = \sqrt{u^2 + v^2}$ in which, surge (u) and sway (v) are the linear velocity components over x_b and y_b directions, respectively. C_D is the dimensionless drag coefficient, C_L is the dimensionless lift coefficient and C_M is the dimensionless moment coefficient and sgn(.) denotes the signum function.

Consider Figure 2.1, it is easy to notice that the drag force is the same as the thrust force, and (buoyancy+lift) forces balance the weight force. Following in this manner, the fish is said to be neutrally buoyant. In swimming robots, this principle is used in balancing. To get the forward propulsion, the thrust force must overcome the drag force so that the thrust force generated from the pectoral fins is equal to the forces due to the drag plus added mass.

2.5. Modeling the Pectoral Fins of Swimming Robot

A planer motion of the pectoral fins is described in Figure 2.12. The reference frame $F_E = [XYZ]^T$ refers to the earth-fixed frame, and the body reference frame $F_B = [x_b y_b z_b]^T$ with unit vectors $[i, j, k]$ denotes the body coordinates, in which the origin coincides with the center of buoyancy, where T represents the transpose of the matrix. It is assumed that the body reference frame is positioned at the center of the mass of the swimming robot body. The first step in designing the complete prototype of the swimming robot is to test the efficiency of the pectoral fins.

Assuming that the body's frame coincides with the center of mass, since the design is concerned with only the pectoral fins at this stage, the movement of the body is restricted to a 1-DOF, which is represented by the linear velocity component along the x_b-axis.

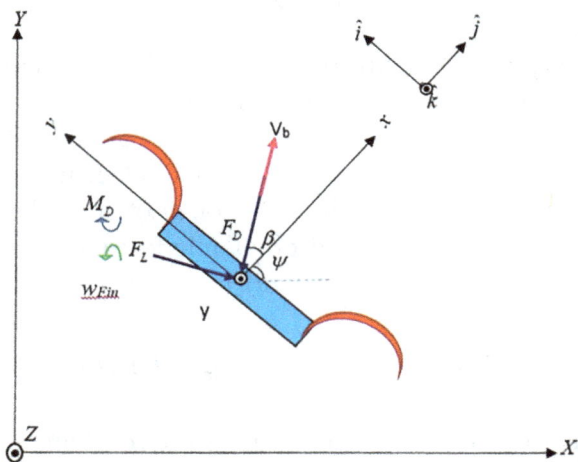

Figure 2.12. Top view of the free body diagram of the proposed robot.

Following the simplified motion equations of the fish body in the FB reference body frame, the robot is assumed to be surrounded by inviscid water and moves in x_b-y_b plane, with the consideration that inertial coupling occurring in the motions of surge (u), sway (v), and yaw (w_{Fin}) is neglected, the dynamic equation in a plane of a rigid body is given by Kirchhoff's equation and written as follows:

$$\left.\begin{array}{l} (M_b - A_{\dot{u}_a})\dot{u} = (M_b - A_{\dot{v}_a})vw_{\text{Fin}} + F_{bx} \\ (M_b - A_{\dot{v}_a})\dot{v} = -(M_b - A_{\dot{u}_a})uw_{\text{Fin}} + F_{by} \\ (I_z - A_{\dot{w}_{\text{Fin}}})\dot{w}_{\text{Fin}} = M_{bz} \end{array}\right\} \quad (2.7)$$

where M_b is a total swimming robot mass (with pectoral fins only), I_z denotes the body's inertia value about the z_b-axis, and $A_{\dot{u}a}$ is the added mass on x_b-axis, $A_{\dot{v}a}$ is the added mass on y_b-axis, and $A_{\dot{w}\text{Fin}}$ is the added inertia on z_b-axis of the robot's body. The variables F_{bx}, F_{bx} and M_{bz} denote the external forces exerted by the suggested body, and described as follows:

$$\left.\begin{array}{l} F_{bx} = F_{hx} - F_D \cos\beta + F_L \sin\beta \\ F_{by} = F_{hy} - F_D \sin\beta - F_L \cos\beta \\ M_{bz} = M_{hz} + M_D \end{array}\right\} \quad (2.8)$$

where F_{hx}, F_{hy} and M_{hz} are the generated hydrodynamic resistant forces and moments transferred from the pectoral fins to the proposed body of the robot. These forces are evaluated based on blade element theory by dividing each fin into N rigid elements as shown in Figure 2.11, each element with radius of dr of equal length G such that $G = PF_L/N$. Then the hydrodynamic force at G over pectoral fin surface area $S = 4\pi dr_{\text{Fin}}^2/(360°/\theta_R)$ is equal to:

$$dF_{hi}(G) = -\frac{1}{2}\rho V_G^2 C_*(4\pi dr_{\text{Fin}}^2/(360°/\theta_R))\hat{e}_{V_G} \quad (2.9)$$

where r_{Fin} is the radius of each fin, θ_R is the revolving angle of each fin, V_G is the velocity of element G at time t, $\hat{e}_{V_G}$ is a unit vector in the direction of V_G and C_* is the force coefficient such that ($*$) represents either lift force coefficient C_L or drag force coefficient C_D such that:

$$C_* = \begin{cases} C_L = 3.4\sin(\theta_G) \\ C_D = 0.4\cos^2(2\theta_G) \end{cases} \qquad (2.10)$$

where θ_G is the hydrodynamic angle of attack of element G. Then the hydrodynamic forces will be found as:

$$F_{hi}(t) = \int_0^G dF_{hi}(G, t) \qquad (2.11)$$

The kinematics of the robot is given as:

$$\left. \begin{aligned} \dot{X} &= u\cos\psi - v\sin\psi \\ \dot{Y} &= v\cos\psi + u\sin\psi \\ \dot{\psi} &= w_{\text{Fin}} \end{aligned} \right\} \qquad (2.12)$$

where ψ represents the angle that lies between the x_b-axis of the body and X-axis of earth reference frame. β refers to the body's angle of attack when swimming in the water. Generally, the drag force is the force that is in the opposite direction to the flow and the lift force is the normal force to the flow.

2.6. Hydrodynamic Force of Pectoral Fins

The influence of resistance of the water against the robot's movement is known as the hydrodynamic force. This hydrodynamic force can be considered as a function of a robot's motion parameter. When an object moves in the water, a portion of the water moves within the movement of the body. This moving water forms an additional body mass in addition to the original body mass. The suggested system design with the shape of the pectoral fin is concave; it will allow a portion of water to move along within the body according to the fin movement.

As mentioned early, there are two strokes of fin movement, the first one is the power stroke which occurs when the fins move toward the back of the body within velocity $V_{\text{Fin(power)}}$, the second is the recovery stroke, which occurs when the fins return back to their initial position within velocity $V_{\text{Fin(recovery)}}$. In order to get thrust, the ratio

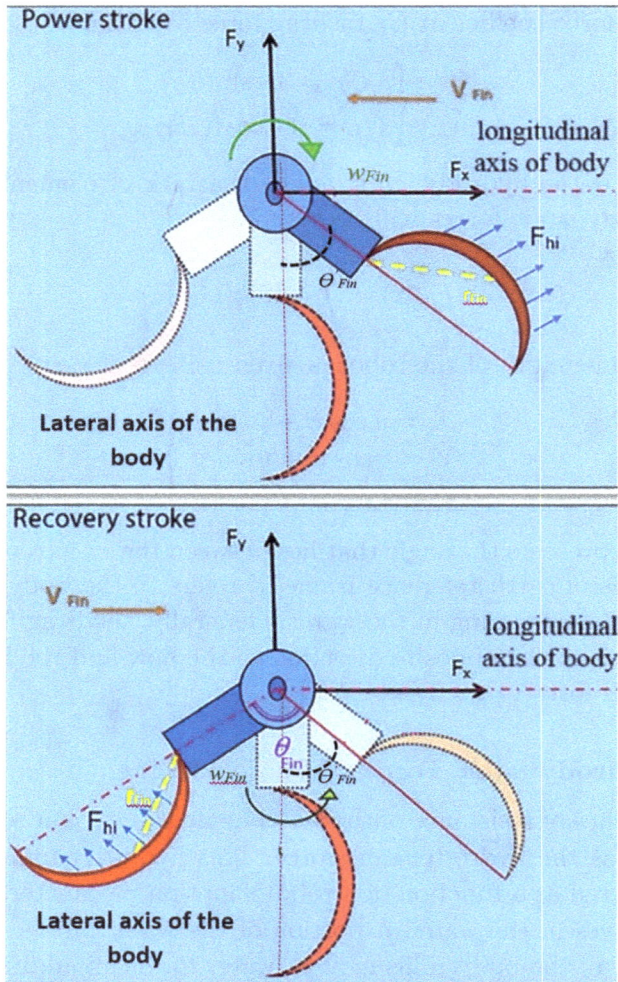

Figure 2.13. Pectoral fin motion during two different strokes.

of $R_F = V_{\text{Fin(power)}}/V_{\text{Fin(recovery)}}$ is varied to obtain the optimum R_F as shown in Figure 2.13.

The blue arrows represent the reaction forces F_{hi} generated as the fin oscillates back and forth, due to fin shape, where the subscript i represents the generated hydrodynamic forces in x_b, y_b and z_b directions, respectively. These forces are at maximum value along

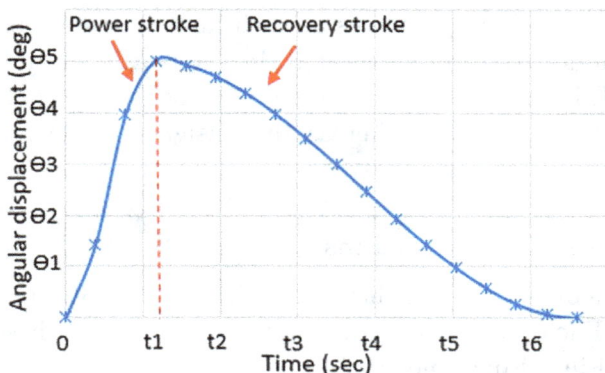

Figure 2.14. Angular position during power and recovery strokes.

the power stroke and minimum value along the recovery stroke, this way the total thrust will be generated.

Figure 2.14 demonstrates the motion input signal during the power stroke and recovery stroke periods. However, the starting angle (θ'_{Fin}) that lies between the fin base and the lateral axis of the body is also a parameter that affects the hydrodynamic force generation. The pectoral are fins placed in an angular position $\theta_{\text{Fin}}(t)$ such that:

$$\theta_{\text{Fin}}(t) = (A/2) - (A/2) \cos w_f t \qquad (2.13)$$

where $\theta_{\text{Fin}}(t)$ represents the instantaneous angular position of the base of the fin, A is the amplitude of the wave generated by the fin, and w_f is the angular frequency which can be given as $w_f = 2\pi f$.

Only the forward direction of swimming is considered. The main goal, at the current stage, is to determine the hydrodynamic forces applied to the pectoral fin while moving toward or backward within different fin beat angles. In order to get net thrust, the following fact holds:

$$\text{Net}_{\text{Thrust}} = \sum_{\theta_{\text{Fin}}}^{-\theta_{\text{Fin}}} \text{Thrust} - \sum_{-\theta_{\text{Fin}}}^{\theta_{\text{Fin}}} \text{Drag} \qquad (2.14)$$

where the hydrodynamic forces at each angle of rotation are calculated and added to the previous angle in a step of $10°$ until completing one cycle. It is noteworthy that the sign $(-)$ before the angle indicates the recovery stroke side, whereas the unsigned angle refers to the starting angle at the power stroke.

2.7. Design of Pectoral Fins

Three types of pectoral fins have been designed by SOLIDWORKS® platform. Each one has a different surface area; these fins are named as Fin1, Fin2, and Fin3. The specifications of these fins are listed in Table 2.1, each fin has the same length $(PF_L = 5\,\text{cm})$ but with different revolution angles about its diameter, which results in different Ar as shown in Figure 2.15. A 3D printer has been used to print the pectoral fins and joints by using PLA material (Polylactic Acid) of the density as $1240\ \text{kg/m}^3$ as in Figure 2.16. Each fin is attached to a pair of servomotors via very precisely designed joints as shown in Figure 2.17. Two joints were attached to each servomotor to

Table 2.1. Pectoral fins specifications.

Fin type	Outer radius	Inner radius	Geometrical revolving angle θ_R	Projected surface area S	Mass of each fin
Fin1	2.5 cm	2.3 cm	90°	16.6 cm²	5.00 gm
Fin2	2.5 cm	2.3 cm	45°	8.31 cm²	2.82 gm
Fin3	2.5 cm	2.3 cm	22.5°	4.15 cm²	1.76 gm

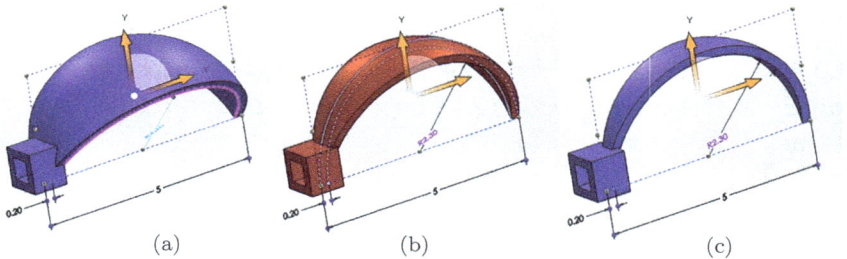

Figure 2.15. Three different concave pectoral fins. (a) Fin1. (b) Fin2. (c) Fin3.

Figure 2.16. Real size of the pectoral fin.

Figure 2.17. Pectoral fin, joint and servomotor designs in SOLIDWORKS®.

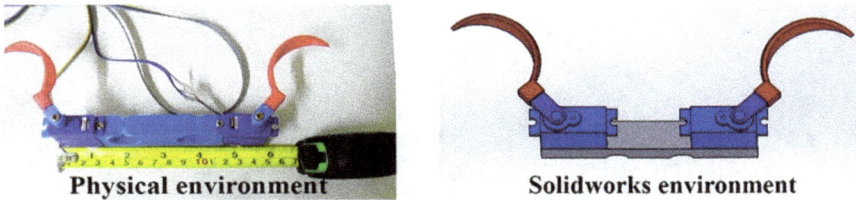

Physical environment   **Solidworks environment**

Figure 2.18. Physical and SOLIDWORKS® designs for the pectoral fins system.

satisfy the motion dictated by the servomotor such that the pectoral fin maintains the highest speed during the power stroke and the lowest speed during the recovery stroke.

The servomotors are linked with pectoral fins through their arms and then placed on a thin plate, which in turn slides over two parallel shafts. The shafts are fixed in a swimming pool of acrylic plastic material of 1 m in length, 0.65 m in both width and height. The complete prototype design of the pectoral fins experiment is shown in Figure 2.18.

2.8. Validation of Pectoral Fins Design

The designed 3D model of the swimming robot is implemented with a computational domain of 1 m, 0.65 m, and 0.65 m for length, width, and height, respectively. This domain has been used to match the dimension of the physical swimming pool as shown in Figures 2.19 and 2.20.

The physical model of the swimming pool is made from acrylic plastic material. Two waterproof digital servomotors of 3 Kg-cm from Hitec (HS-5086WP) have been used. The Atmega microcontroller has been employed to control the servomotors motions via wires. A 3D printer of PLA material has been used to print all plastic parts such as the robot body, fins, base plate and joints, as shown in Figure 2.20. A high-speed camera from Kodak is used, the camera has high resolution of 1920×1080 pixels and 30 frame rate per second (fps) and was positioned at 50 cm from the top view of the robot to capture the forward swimming velocity of the robot. The recorded videos are analyzed by computer vision techniques within MATLAB platform to extract the desired data and process the video via the

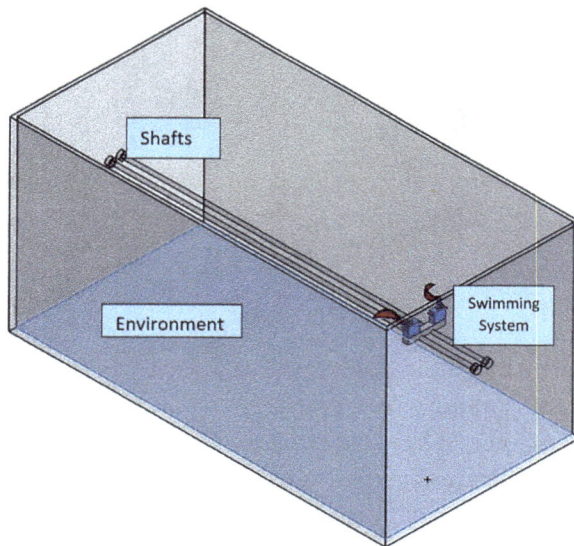

Figure 2.19. The swimming pool by SOLIDWORKS®.

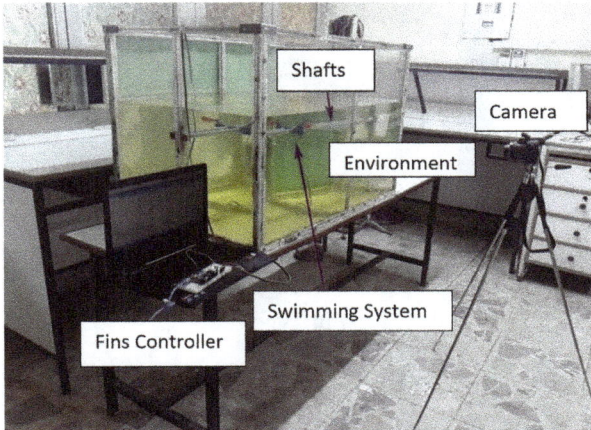

Figure 2.20. The physical model of the swimming pool.

image processing toolbox. All electronic devices are covered with NANO PROTECH coating technology spray to protect them from water contact.

The term "swimming robot" used in this chapter, refers to (plate, servomotors, joints, and pectoral fins). In the next chapters, it is used to denote the complete body design of the robot.

2.8.1. *The Computational Fluid Dynamics (CFD) Analysis*

The aspect ratio Ar of each distinct fin provided in Figure 2.15 is calculated as shown in Figure 2.21. The three types are analyzed with the CFD method; a computational domain is set to fit the size of each fin as shown in Figure 2.22. To show some realistic results, flow type has been set as (laminar and turbulent) with a static pressure of 101325 Pa at 293.2 K and a local mesh of six levels of refinement cells is used throughout this simulation.

Since rowing motion in Labriform mode is "drag-based", the generated hydrodynamic forces (thrust) should be at maximum during the power stroke (i.e. when the fin starts to hit) and minimum during recovery stroke (i.e. when the fins are returned to its initial position). Drag, lift and moment coefficients depend on the angle

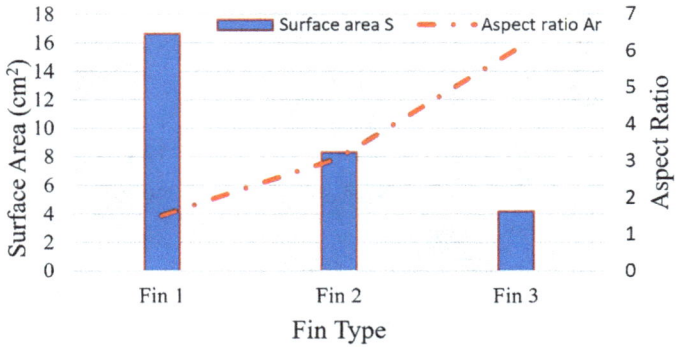

Figure 2.21. Aspect ratio of each fin.

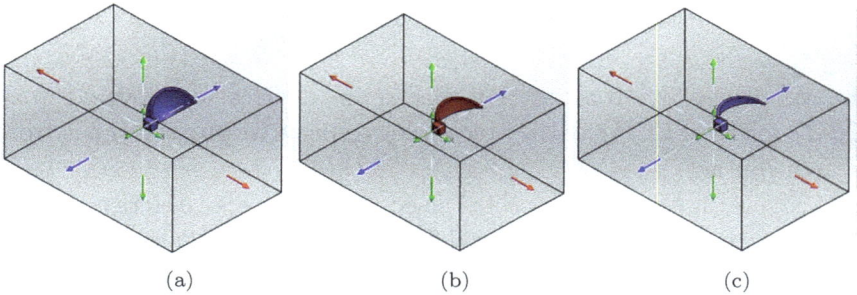

Figure 2.22. CFD domains for each fin. (a) Fin1. (b) Fin2. (c) Fin3.

of attack and Reynolds number. The application under study is of order 10^3 of Reynolds number. These hydrodynamic coefficients can be determined directly via CFD simulations or by experiments. Since the application under study is restricted to 1-DOF represented by surge velocity, then the hydrodynamic forces generated will be considered in this direction only.

At power stroke time, each fin is subjected to a flow of V_b in the x_b-direction. Flow trajectories provided in Figure 2.23 show that Fin1 with the lowest aspect ratio Ar as expressed in equation (2.4), has the biggest surface area and hence the biggest drag force generation. In contrast, for the case of Fin3, which has the highest aspect ratio Ar, although it produces the lowest drag, but the total generated thrust is not sufficient to accelerate the proposed design as tested

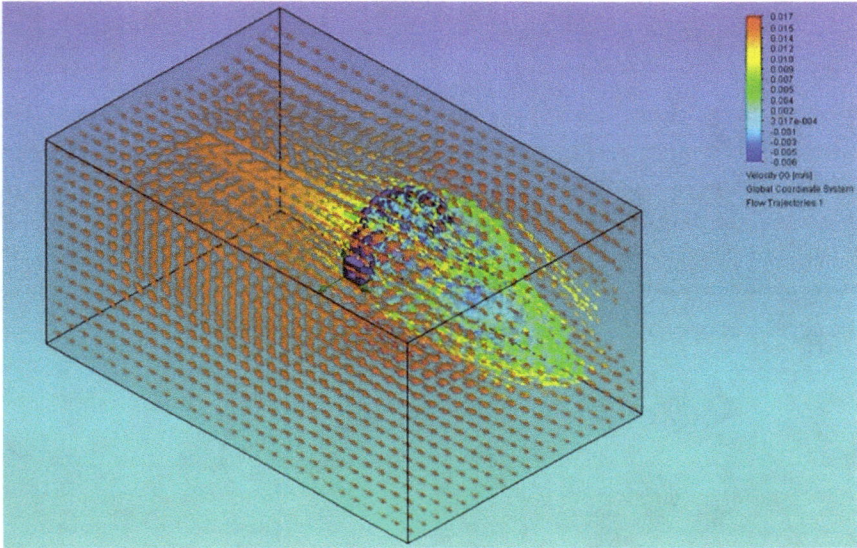

(a)

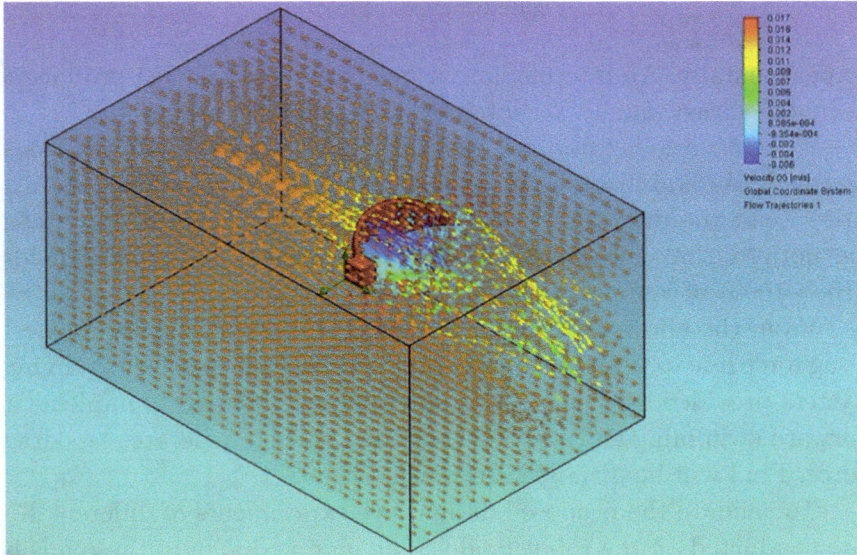

(b)

Figure 2.23. Velocity trajectories during power stroke. (a) Fin1. (b) Fin2. (c) Fin3.

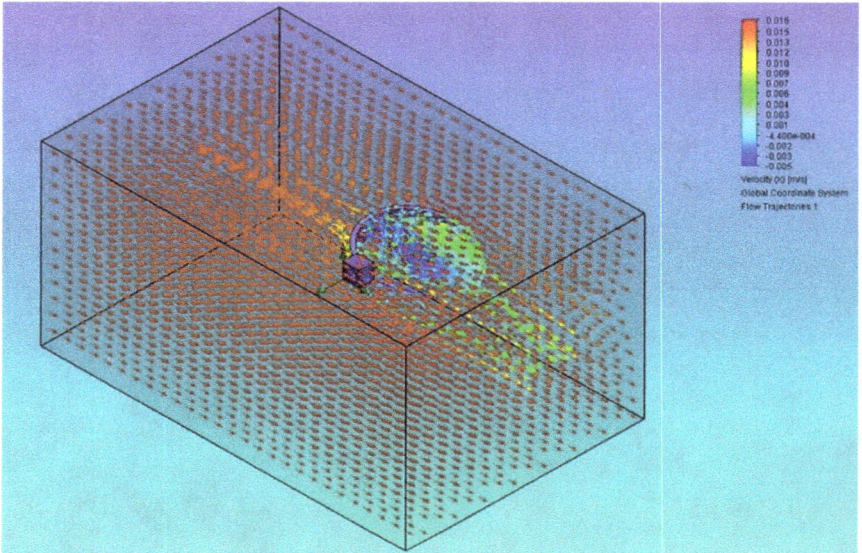

(c)

Figure 2.23. (*Continued*)

experimentally. An intermediate aspect ratio of Fin2 will produce a valuable thrust and an accepted drag at different velocities.

This scenario is examined with CFD analysis, with 300 iterations to ensure the stabilization of the final values as in Figure 2.24. Results show both the drag forces and drag coefficients for each fin. It can be noticed that the drag coefficient is constant at high Reynolds (Re > 1000) of laminar flow.

From the above analysis, it is obvious that Fin1 has the highest drag force due to the large surface area so it will be used for the diving system provided in Chapter 5. While Fin2, will be used to implement forward swimming and steering processes, since it generates less drag force. Fin3 will be dismissed from this study.

To validate the proposed design, five experiments of different R_F are conducted on the robot model. The unique design of these fins is precisely built as an octal hollow spherical shape such that the outer radius is 2.5 cm and the inner radius is 2.3 cm, producing a rigid concave shape of thickness 2 mm. A joint of length 1.88 cm is driven by a servomotor, and attached to each pectoral fin.

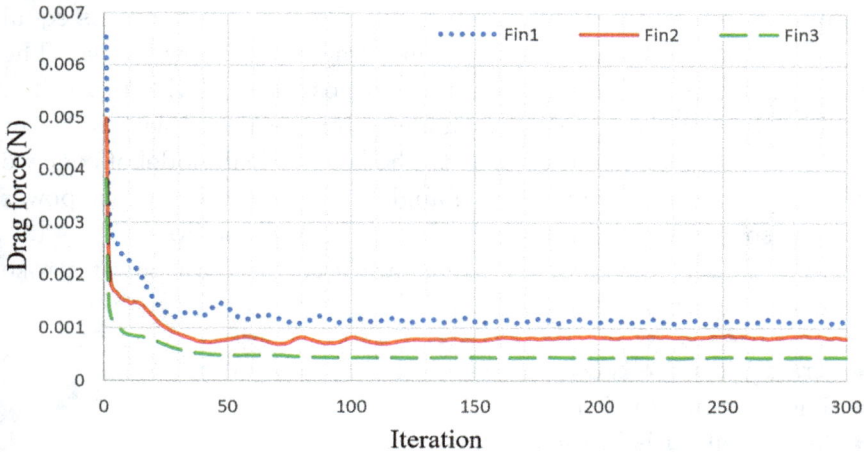

(a)

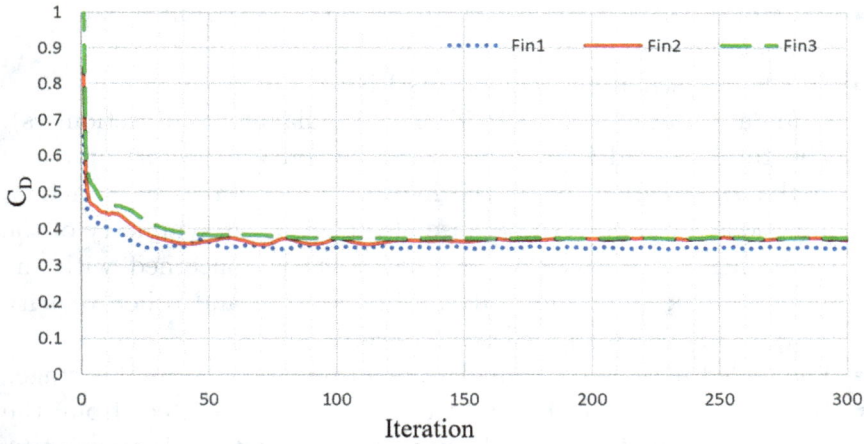

(b)

Figure 2.24. The total drag forces and drag coefficients exerted by Fins. (a) Drag force for each fin. (b) Drag coefficient for each fin.

Since the proposed model is based on rowing motion of the Labriform mode, during the power stroke phase, the thrust force should be at a maximum value, whereas in the recovery stroke phase, the drag force should be minimized as much as possible to generate a net forward thrust. By utilizing the concave shape of the pectoral fin the maximum thrust is produced during the power stroke and

minimum drag force during recovery stroke. The net thrust is equal
to the thrust at power stroke minus drag at recovery stroke. The
model is tested with five different R_F ratios (i.e. of 1:1, 2:1, 3:1, 4:1,
and 5:1), and in each ratio, the model is further tested with different
oscillating amplitude range. Due to the geometrical model of concave
fin design, the maximum starting angle is $\theta'_{Fin} = 50°$ during the power
stroke, so the fin will not hit the frontal part of the body.

The following parameter values are used for both power stroke
and recovery stroke:

- Gravity feature is considered, where $g = 9.81 \, \text{m/s}^2$.
- The roughness is $0 \, \mu\text{m}$.
- No cavitation is in the simulation.
- Water density is $1000 \, \text{kg/m}^3$.
- Plate of the servos' dimensions is $0.1 \times 0.12 \times 0.01 \, \text{m}$.
- Servomotor speed is $0.11 \, \text{sec}/60°$.
- Maximum oscillating frequency is $1.515 \, \text{Hz}$.
- Boundary condition is set to ideal wall (adiabatic and frictionless).
- The total hydrodynamic forces exerted on the pectoral fins are
  in the x_b direction, while the hydrodynamic forces in y_b and z_b
  directions are both zero because of the left-right symmetry of the
  oscillating fins. Consequently, the model is concerned with the
  forward velocity of the body at the x_b-axis, and other velocity
  components on y_b- and z_b-axes are zeros.
- The added mass is very small and can be neglected at this stage.
- To verify the total hydrodynamic forces generated from the
  concave design of the fins, the weight of the plate and servomotors
  is ignored.

2.8.2. *Simulation Results*

The thrust and drag are investigated for every angle of rotation
starting from $50°, 40°, 30°, 20°, 10°, 0°, -10°, -20°, -30°, -40°$, and
$-50°$. Each angle is tested for both power stroke and recovery stroke.
In each case, as shown in Figure 2.25, there is an optimal range of
oscillation with optimal R_F in which the total accumulated drag force
is minimum. For the ratio of 1:1, the generated drag forces are given

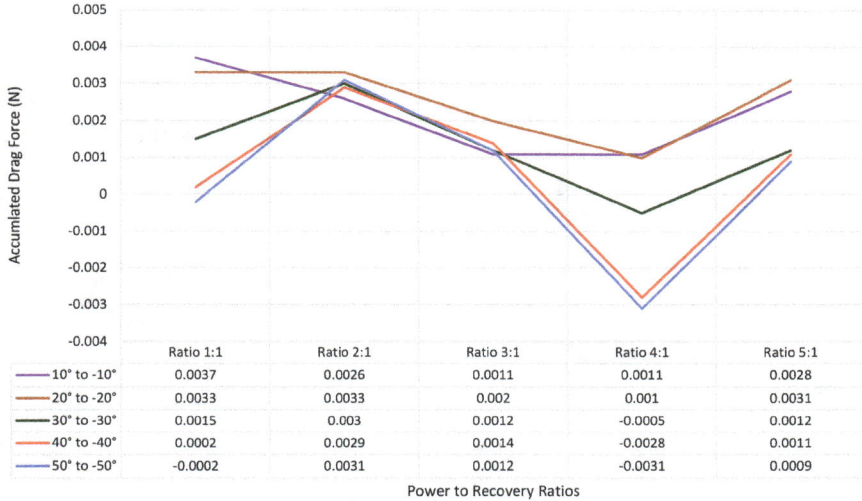

Figure 2.25. Hydrodynamic drag forces vs different ranges of R_F.

for the different amplitude ranges. Starting from $(10°$ to $-10°)$, it can be noticed that the drag force is equal to 0.0037 N as shown in Figure 2.25.

In case of higher ranges of amplitude used, this value is added to the lower range force. The total accumulated drag at $R_F = 1$ at $(50°$ to $-50°)$ is equal to $(0.0037 + 0.0033 + 0.0015 + 0.0002 + 0.0002)$ which is equal to 8.9 mN. In the same manner, the total drag forces are 14.9 mN, 6.9 mN, 8.5 mN, and 9.1 mN, for R_F 2, 3, 4, and 5, respectively. Therefore, the moderate ratio of $R_F = 3$ can be considered as an optimal with approximately all different ranges. It can also noticed that some ranges give a negative drag, however this sign refers to the direction of water flow reversed in the other direction. As the range increases, the hydrodynamic thrust force will be increased, this fact is validated as in Figures 2.26 to 2.30. The generation of the hydrodynamic forces is studied in detail in the next chapter.

The results show that, with the increasing of amplitude of the angular displacement of the pectoral fins, the amount of forward thrust increases as shown in Figure 2.31. The maximum thrust can be

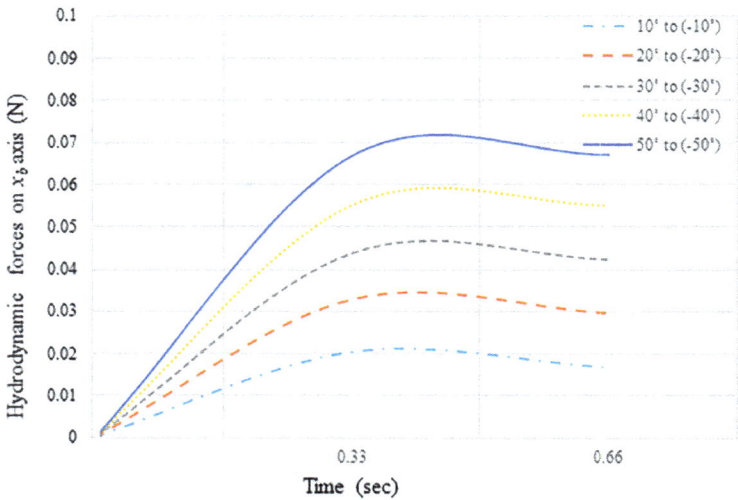

Figure 2.26. Hydrodynamic forces on x_b-axis (N) when R_F is 1:1.

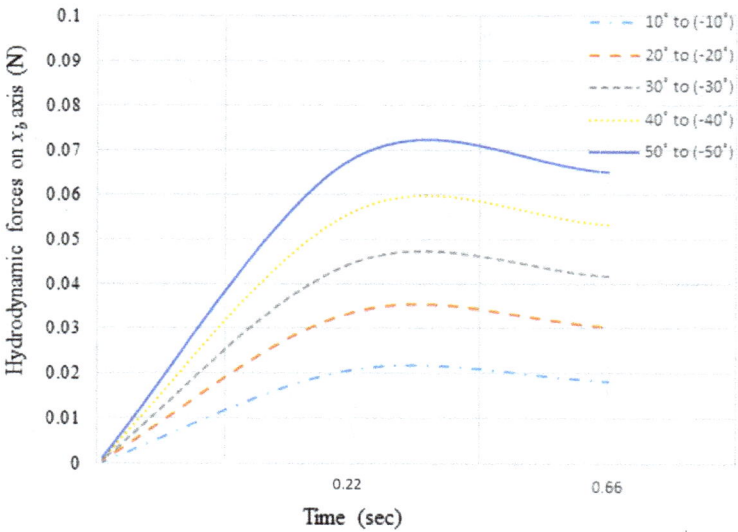

Figure 2.27. Hydrodynamic forces on x_b-axis (N) when R_F is 2:1.

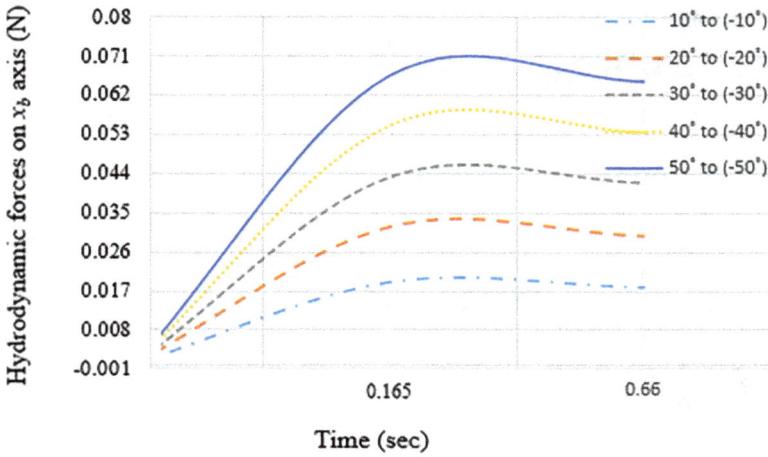

Figure 2.28. Hydrodynamic forces on x_b-axis (N) when R_F is 3:1.

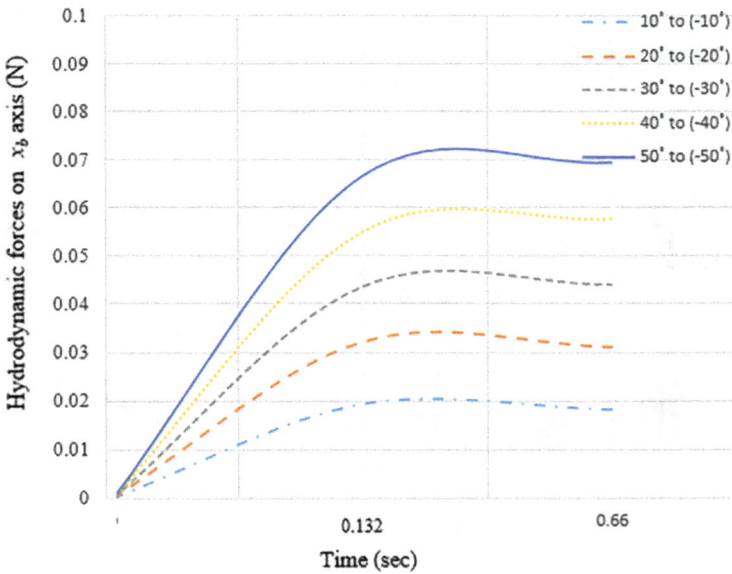

Figure 2.29. Hydrodynamic forces on x_b-axis (N) when R_F is 4:1.

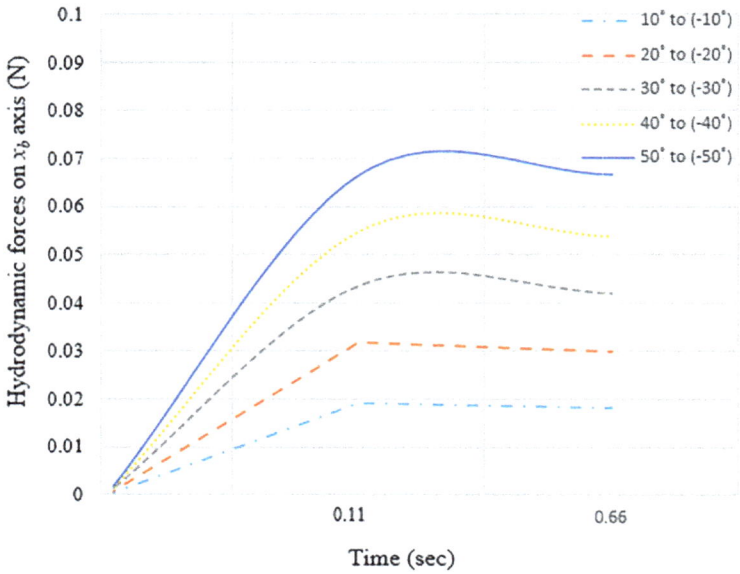

Figure 2.30. Hydrodynamic forces on x_b-axis (N) when R_F is 5:1.

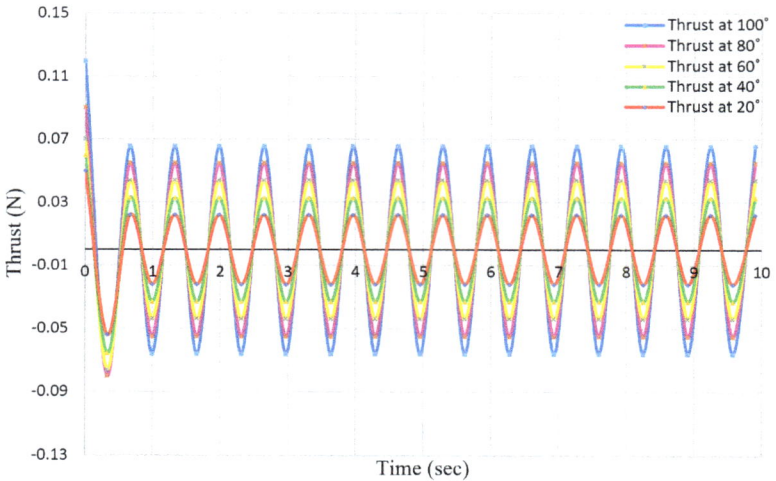

Figure 2.31. Total thrust over different oscillating amplitude ranges.

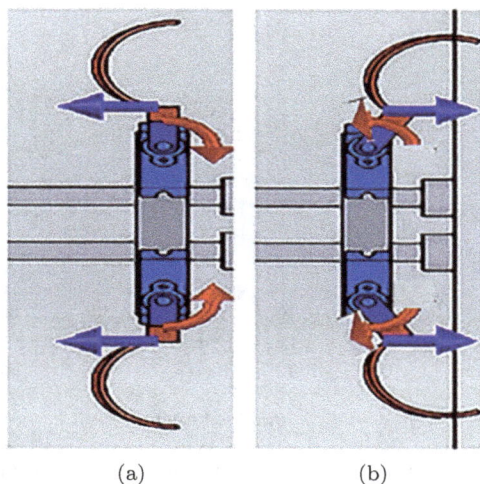

Figure 2.32. Thrust and drag force: (a) Power stroke. (b) Recovery stroke.

achieved at 50° and the minimum drag is at −50°, producing a total angular displacement of 100° around the y_b-axis. The hydrodynamic forces are applied to both concave and convex sides when the fin starts to beat in power stroke cycle. For a power to recovery stroke ratio of 1:1, the active force is the forward thrust force and is applied from 0 to 0.33 sec. In recovery stroke, the active force is the drag force, which is in the opposite direction of robot's forward movement and is applied from 0.33 to 0.66 sec to complete one cycle, as shown in Figure 2.32. The red arrows indicate the motor motion and the blue ones represent the applied hydrodynamic forces. For other ratios, the same manner is followed, where each time corresponds to a specific R_F.

The drag coefficient is calculated as shown in Figure 2.33 at R_F of 2:1 at different angles of rotation at Reynolds number $(10^4 - 10^6)$.

For theoretical verifications, the ratio of 1:1 is considered with an oscillating range of 50° to −50°. The angular velocity of the fin at power stroke speed at time 0.33 sec is about 303°/sec; this value matches the theoretical calculations by the relation $w_{\text{Fin}} = \theta_{\text{Fin}}/t$, where θ_{Fin} is 100° (1.74 rad/sec) and t is 0.33 sec, which gives the angular velocity of 5.27 rad/sec (301.99°). Figure 2.34 shows the simulation result of angular fin velocity.

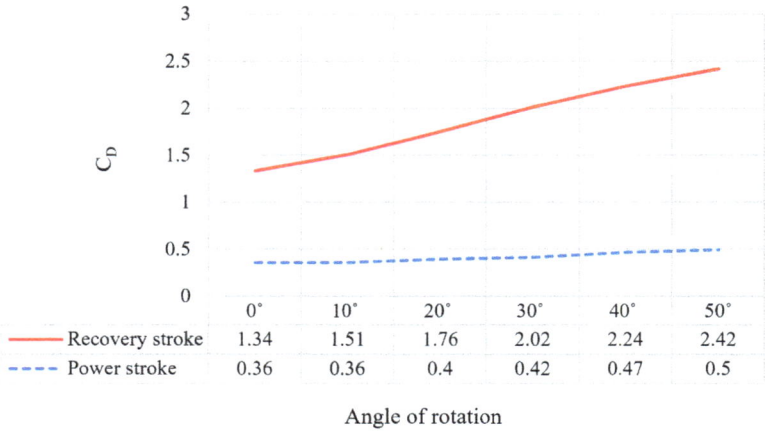

	0°	10°	20°	30°	40°	50°
Recovery stroke	1.34	1.51	1.76	2.02	2.24	2.42
Power stroke	0.36	0.36	0.4	0.42	0.47	0.5

Angle of rotation

Figure 2.33. Drag coefficient at power and recovery strokes.

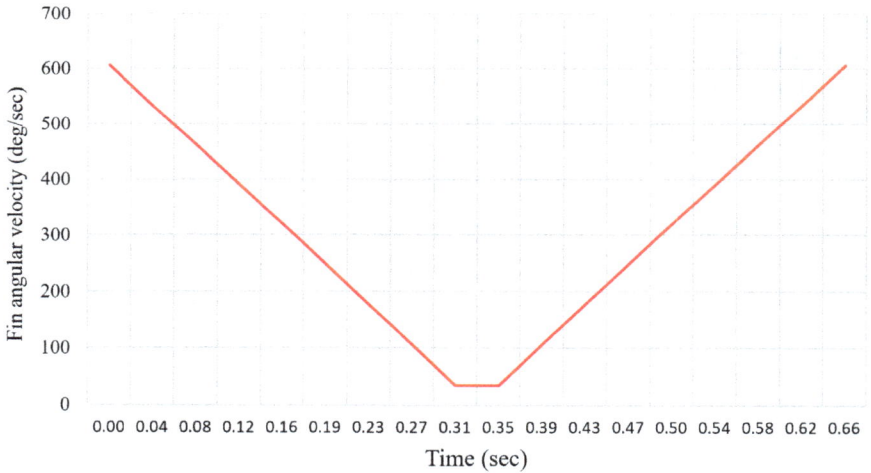

Figure 2.34. Pectoral fin angular velocity at R_F 1:1.

The corresponding body velocity is then calculated at different values of R_F as in Figure 2.35. The body velocity reaches about 5 to 10 cm/sec. The required torque of the motor to actuate the pectoral fin is calculated with and without the hydrodynamic force as shown in Figure 2.36. The results match the theoretical calculation with

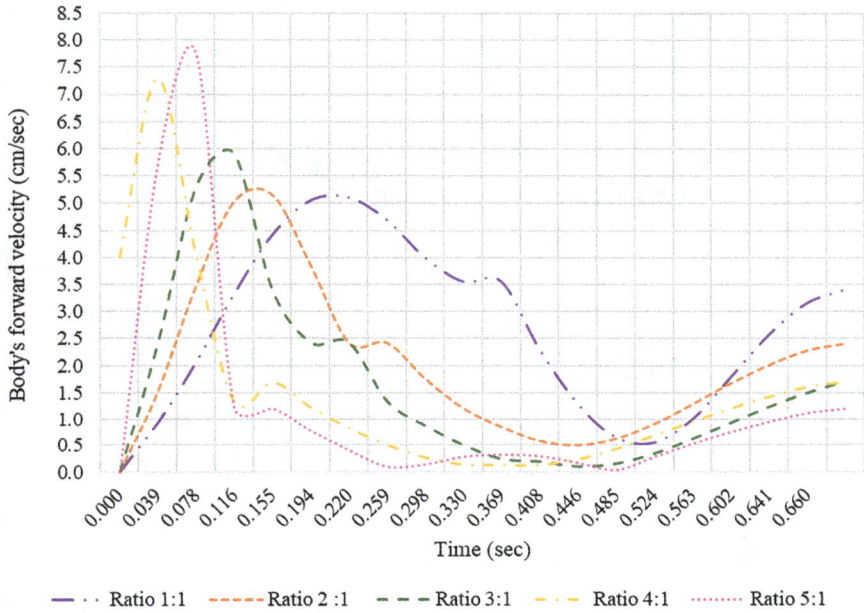

Figure 2.35. Forward body velocity with different power to recovery ratios.

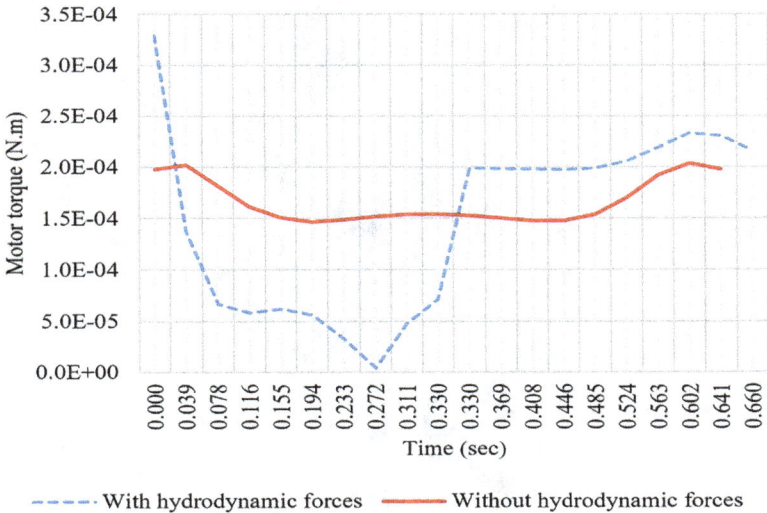

Figure 2.36. Motor torque at power and recovery strokes.

motor torque equal to the applied force multiplied by radius (i.e. $\approx 0.065 \times 0.05 \approx 3 \times 10^{-3}$); since the applied force is considered only on the fins without taking into account the other components, the motor torque is very small at this stage.

2.8.3. *Practical Experiments Result*

Two stainless steel parallel shafts used as motion straighter are fixed in the middle of the tank at a height of 32.5 cm; about half of the pool is filled with water to make the robot neutrally buoyant. The robot is driven by wires, which are connected to the 5V power supply. Figure 2.37 shows a top view of the complete setup process of the experiment. Five R_F ratios were tested. In all experiments, the maximum value of servomotor speed is fixed, which corresponds to a fin beating frequency of 1.515 Hz.

A ratio of 1:1 is experimented and compared to the simulated results as shown in Figures 2.38 and 2.39. Snapshots in Figure 2.38 show the time corresponding to the movement of the fins, where the

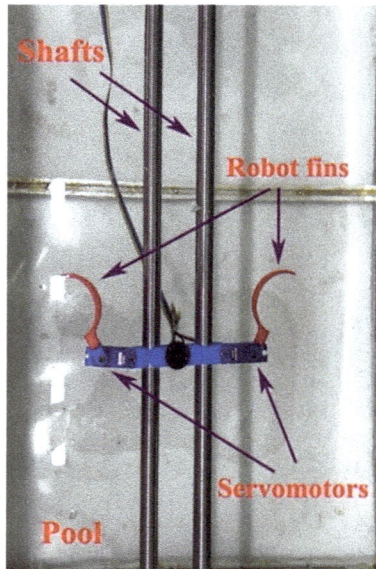

Figure 2.37. Top view of the robot in the pool.

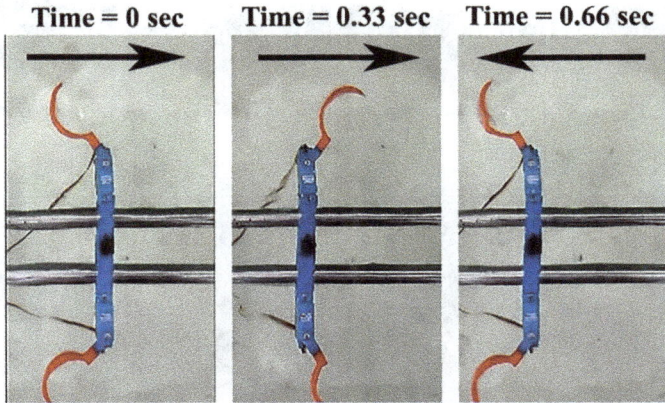

Time = 0 sec   Time = 0.33 sec   Time = 0.66 sec

Figure 2.38. Snapshot of complete one cycle at R_F of 1:1.

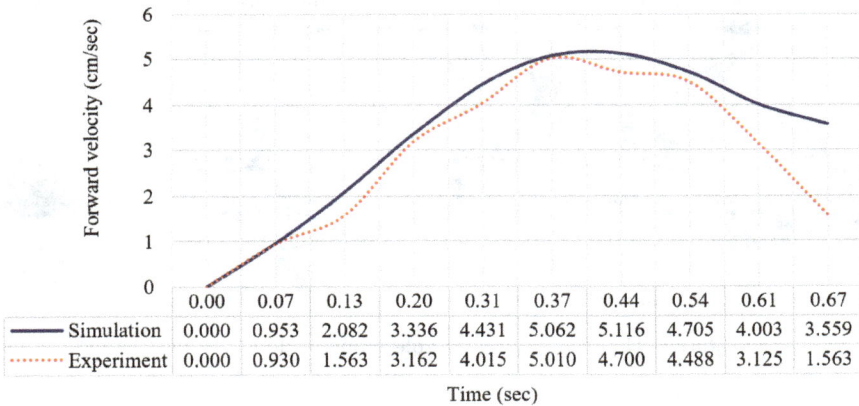

	0.00	0.07	0.13	0.20	0.31	0.37	0.44	0.54	0.61	0.67
Simulation	0.000	0.953	2.082	3.336	4.431	5.062	5.116	4.705	4.003	3.559
Experiment	0.000	0.930	1.563	3.162	4.015	5.010	4.700	4.488	3.125	1.563

Time (sec)

Figure 2.39. Forward velocity when the power to stroke ratio is 1:1.

time duration for the servomotor to complete one cycle is 0.66 sec. Therefore, it can be noticed that the time taken at the power stroke is the same as the time taken at the recovery stroke time. The gained thrust is very small at this ratio.

On the other hand, in Figure 2.40, the ratio is set to 2:1 where the fin completes the power stroke at time 0.22 sec, while the recovery stroke will be completed at time 0.66 sec, which is twice of the time taken to complete the power stroke.

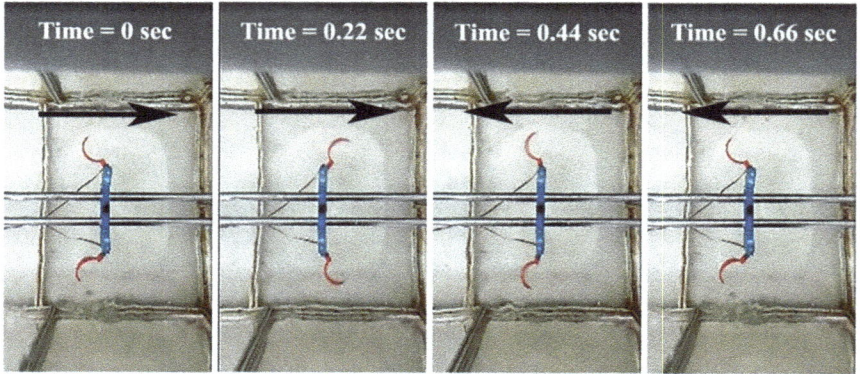

Figure 2.40. Snapshot of complete one cycle at R_F of 2:1.

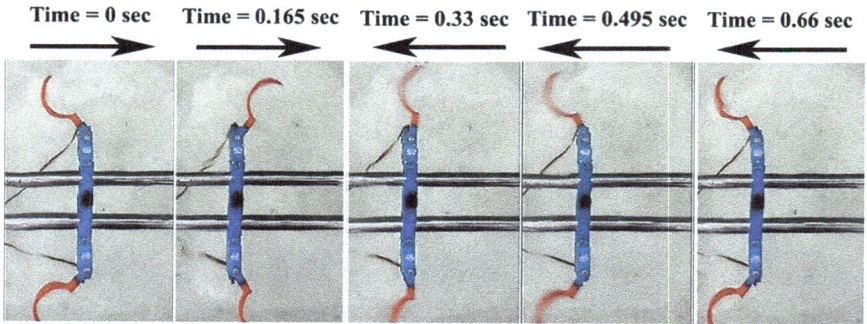

Figure 2.41. Snapshot of complete one cycle at R_F of 3:1.

In Figure 2.41, the power stroke will be completed at time 0.165 sec, and from 0.165 to 0.66 sec, the fins will complete the recovery stroke; the time taken to complete the recovery stroke is three times the time taken to complete the power stroke.

In Figure 2.42, the velocity of the power stroke is further increased and will be completed at time 0.132 sec, while the remaining time will be taken by the recovery stroke to be completed at time 0.66 sec.

Finally, in Figure 2.43, the velocity of power stroke is now five times the velocity of the recovery stroke, the power stroke will be

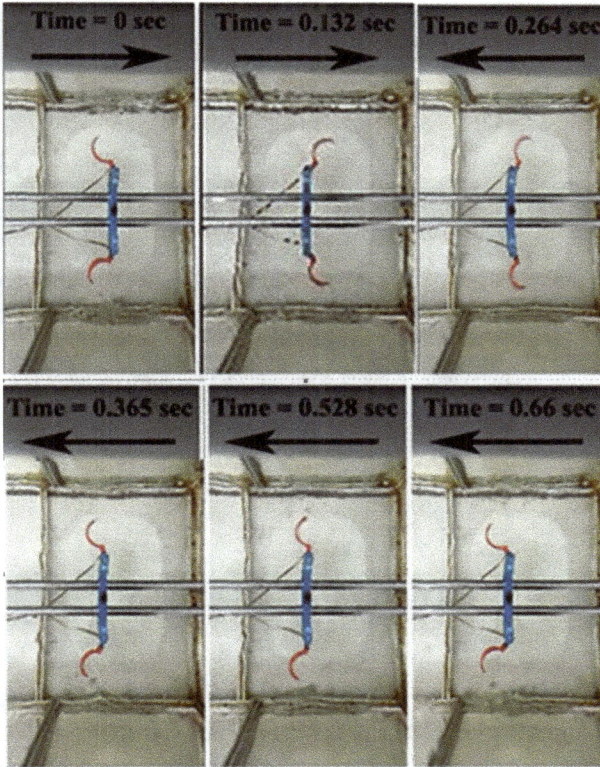

Figure 2.42. Snapshot of complete one cycle at R_F of 4:1.

completed at time 0.11 sec, and time from 0.11 to 0.66 sec will be for the recovery stroke.

For all the designed rigid fins, it can be noticed that as the ratio of R_F increases, at $R_F = 1$ and $R_F = 2$, it provides a swimming velocity that matches the results shown in Figure 2.35. Since the drag force will be increased at higher R_F within the servomotor constraints, it can be noticed that the highest achievable swimming velocity does not lie at the highest R_F. Instead of that, the highest swimming velocity is obtained if intermediate ratio of 3:1 is used. When the ratio is 3:1, the drag force at the recovery stroke will no longer affect the forward velocity of the body. Figures 2.39 and 2.44–2.47 demonstrate this fact.

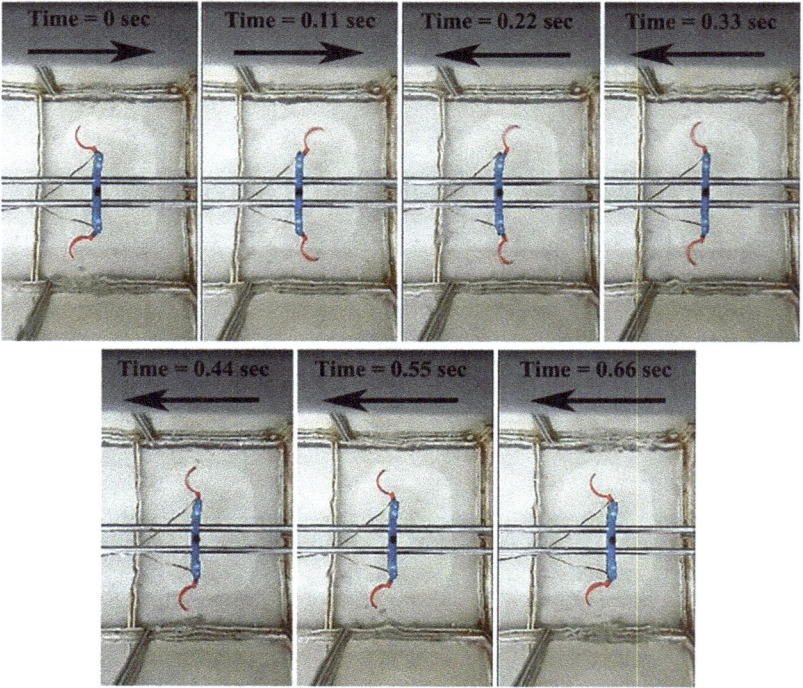

Figure 2.43. Snapshot of complete one cycle at R_F of 5:1.

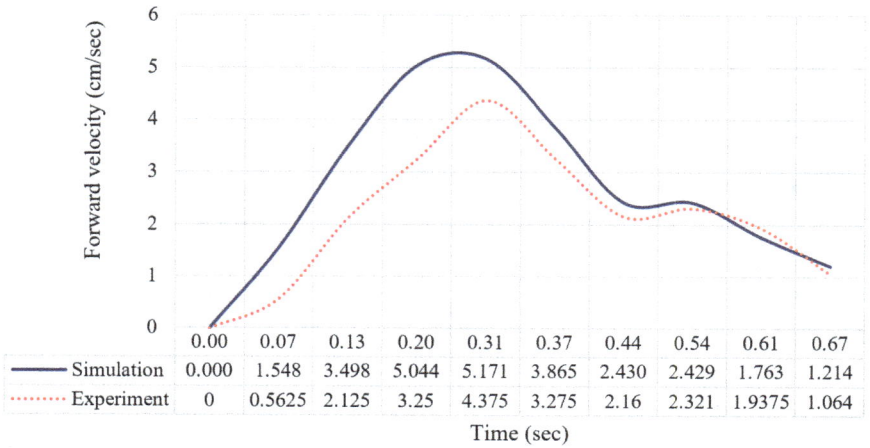

	0.00	0.07	0.13	0.20	0.31	0.37	0.44	0.54	0.61	0.67
Simulation	0.000	1.548	3.498	5.044	5.171	3.865	2.430	2.429	1.763	1.214
Experiment	0	0.5625	2.125	3.25	4.375	3.275	2.16	2.321	1.9375	1.064

Time (sec)

Figure 2.44. Forward velocity when R_F is 2:1.

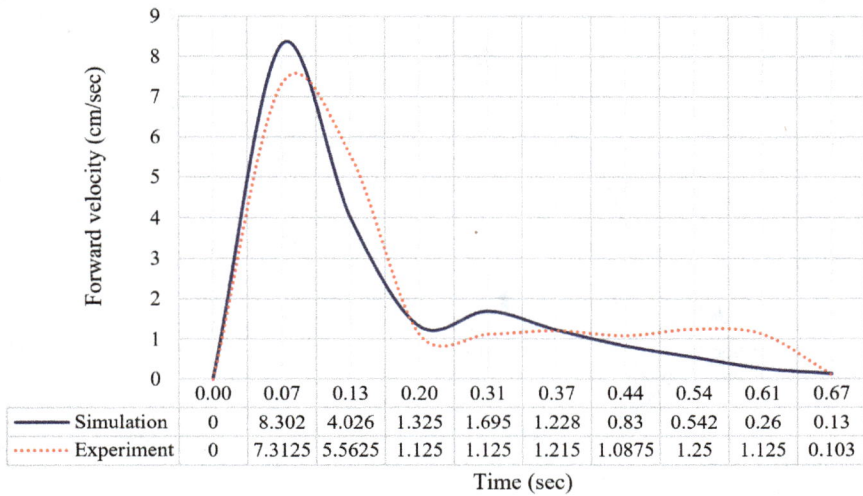

Time (sec)	0.00	0.07	0.13	0.20	0.31	0.37	0.44	0.54	0.61	0.67
Simulation	0	8.302	4.026	1.325	1.695	1.228	0.83	0.542	0.26	0.13
Experiment	0	7.3125	5.5625	1.125	1.125	1.215	1.0875	1.25	1.125	0.103

Figure 2.45. Forward velocity when R_F is 3:1.

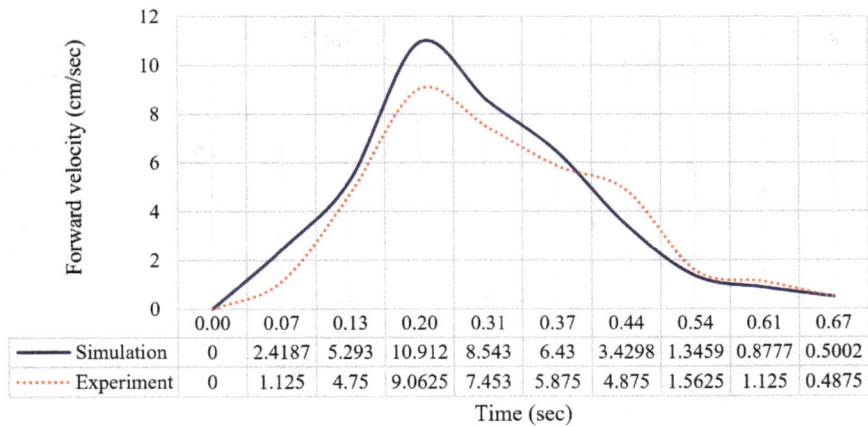

Time (sec)	0.00	0.07	0.13	0.20	0.31	0.37	0.44	0.54	0.61	0.67
Simulation	0	2.4187	5.293	10.912	8.543	6.43	3.4298	1.3459	0.8777	0.5002
Experiment	0	1.125	4.75	9.0625	7.453	5.875	4.875	1.5625	1.125	0.4875

Figure 2.46. Forward velocity when R_F is 4:1.

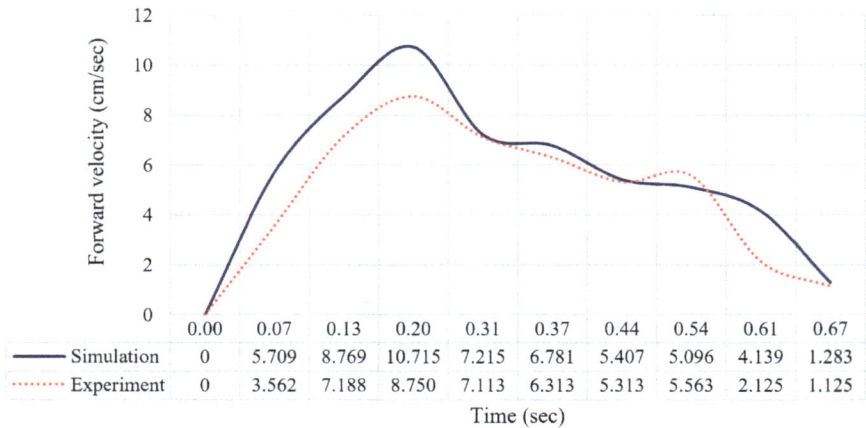

Figure 2.47. Forward velocity when R_F is 5:1.

2.9. Conclusions

The works provided within this chapter represent the first step in designing a Labriform-swimming robot with pectoral fins only. The effect of concave pectoral fins in producing the highest thrust while keeping the drag force to a minimum has been studied carefully. Within the scope of CFD analysis, the optimum shape and size from three distinct concave pectoral fins have been discussed, each one with a specific aspect ratio Ar.

The results showed that the lowest aspect ratio fin (Fin1) produces the highest drag, whereas the highest aspect ratio fin (Fin3) gives the lowest drag. However, an intermediate aspect ratio fin (Fin2) produces a valuable thrust and an accepted drag. Two pectoral fins of (Fin2) size have been attached to two servomotors and fixed over a plastic plate. Utilizing the shape of concave fins, the hydrodynamic forces generated at the power stroke overrides the one generated at the recovery stroke. This fact gives the robot the ability to move forward at larger displacement compared with the recovery stroke.

Several fin-oscillating angles were investigated, and a maximum oscillation range was investigated for each R_F. Five values of R_F have been tested, the generated hydrodynamic thrust and drag are calculated, and the maximum thrust was obtained in the moderate ratio of R_F of 3:1. The drag force will be increasing at the higher ratios, and it can be noticed that the highest achievable swimming velocity does not occur at the highest R_F. Instead of that, the highest swimming velocity is achieved if a moderate ratio (3:1) is used.

References

Behbahani, S. B. and Ta, X. (2016). Bio-inspired flexible joints with passive feathering for robotic fish pectoral fins, *Bioinspiration & Biomimetics*, 11(3).

Behbahani, S. B. and Tan, X. (2016). Design and modeling of flexible passive rowing joint for robotic fish pectoral fins, *IEEE Trans Robot.*, 32, pp. 1119–1132.

Behbahani, S. B. and Tan, X. (2017). Role of pectoral fin flexibility in robotic fish performance, *J. Nonlinear Sci.*, 27(4), pp. 1155–1181.

Bellwood, D. R. and Wainwright, P. C. (2001). Locomotion in labrid fishes: implications for habitat use and cross-shelf biogeography on the Great Barrier Reef, *Coral Reefs*, 20, pp. 139–150.

Bi, S., Ma, H., Cai, Y., Niu, C. and Wang, Y. (2014). Dynamic modeling of a flexible oscillating pectoral fin for robotic fish, *Ind Robot*, 41(5), pp. 421–428.

Blake, R. W. (1977). On ostraciiform locomotion, *J. Mar. Biol. Ass. U.K.*, 57, pp. 1047–1055.

Blake, R. W. (1979). The swimming of the mandarin fish Synchropus picturatus (Callionyiidae: Teleostei), *J. Marine Biol. Assoc. U.K.*, 59, pp. 421–428.

Breder, C. M. (1924). Respiration of factor in locomotion of fishes, *The American Naturalist*, 58(655), pp. 145–155.

Chen, Z. (2017). A review on robotic fish enabled by ionic polymer–metal composite artificial muscles, *Robotics and biomimetic*.

Combes, S. A. and Daniel, T. L. (2001). Shape, flapping and flexion: wing and fin design for forward flight, *Experimental Biology*, 204, pp. 2073–2085.

Eloy, C. (2012). Optimal Strouhal number for swimming animals, *J. Fluids Struct.*, 30, pp. 205–218.

Fontanella, J. E., Fish, F. E., Barchi, E. I., Campbell-Malone, R., Nichols, R. H., Di-Nenno, N. K. and Beneski, J. T. (2013). Two- and three-dimensional geometries of batoids in relation to locomotor mode, *Journal of Experimental Marine Biology and Ecology*, 446, pp. 273–281.

Fossen, T. I. (2011). *Handbook of Marine Craft Hydrodynamics and Motion Control*, 1st Ed., Norway, John Wiley & Sons Ltd., pp. 122.

Fulton, C. J., Bellwood, D. R. and Wainwright, P. C. (2001). The relationship between swimming ability and habitat use in wrasses (Labridae), *Mar. Biol.*, 139, pp. 25–33.

Gazzola, M., Argentina, M. and Mahadevan, L. (2014). Scaling macroscopic aquatic locomotion, *Nature Physics*, 10(10), pp. 758.

Gibouin, F., Raufaste, C., Bouret, Y. and Argentina, M. (2018). Study of the thrust–drag balance with a swimming robotic fish, *Physics of Fluids*, 30.

Lindsey, C. C. (1978). Form, function and locomotory habits in fish, In *Fish Physiology Vol. VII Locomotion*, edited by W. S. Hoar and D. J. Randall. New York: Academic Press, pp. 1–100.

Marchese, A. D., Onal, C. D. and Rus, D. (2014). Autonomous soft robotic fish capable of escape maneuvers using fluidic elastomer actuators, *Soft Robot*, 1(1).

Naser, F. A. and Rashid, M. T. (2019). Design, modeling, and experimental validation of a concave-shape pectoral fin of labriform-mode swimming robot, *Engineering Reports*, 1(5), pp. 1–17.

Naser, F. A. and Rashid, M. T. (2020). Effect of Reynold number and angle of attack on the hydrodynamic forces generated from a bionic concave pectoral fins, *IOP Conf. Ser.: Mater. Sci. Eng.*, 745, pp. 1–13.

Naser, F. A. and Rashid, M. T. (2020). The influence of concave pectoral fin morphology in the performance of labriform swimming robot, *Iraqi Journal for Electrical and Electronic Engineering*, 16(1), pp. 54–61.

Naser, F. A. and Rashid, M. T. (2021). Enhancement of labriform swimming robot performance based on morphological properties of pectoral fins, *J. Control Autom. Electr. Syst.*, 32, pp. 927–941.

Naser, F. A. and Rashid, M. T. (2021). Design and realization of labriform mode swimming robot based on concave pectoral fins, *Journal of Applied Nonlinear Dynamics*, 10(4), pp. 691–710.

Naser, F. A. and Rashid, M. T. (2021). Implementation of steering process for labriform swimming robot based on differential drive principle, *Journal of Applied Nonlinear Dynamics*, 10(4), pp. 737–753.

Naser, F. A. and Rashid, M. T. (2021). Labriform swimming robot with steering and diving capabilities, *Journal of Intelligent & Robotic Systems*, 103(14), pp. 1–19.

Ngo, V. and McHenry, M. J. (2014). The hydrodynamics of swimming at intermediate Reynolds, *The Journal of Experimental Biol.*, 217, pp. 2740–2751.

Pfeil, S., Katzer, K., Kanan, A., Mersch, J., Zimmermann, M., Kaliske, M. and Gerlach, G. (2020). A biomimetic fish fin-like robot based on textile reinforced silicone, *Micromachines (Basel)*, 11(3), pp. 298.

Quinn, D. B., Lauder, G. V. and Smits, A. J. (2015). Maximizing the efficiency of a flexible propulsor using experimental optimization, *J. Fluid Mech.*, pp. 430–448.

Rashid, M. T. and Rashid, A. T. (2016). Design and implementation of swimming robot based on labriform model, *Al-Sadeq International Conference on Multidisciplinary in IT and Communication Science and Applications (AIC-MITCSA)*, pp. 1–6.

Rashid, M. T. and Rashid, A. T. (2016). Design and implementation of swimming robot based on labriform model, *Al-Sadeq Int. Conf. on Multidisciplinary in IT and Commun. Sci. and App. (AIC-MITCSA), Iraq*.

Rashid, M. T., Naser, F. A. and Mjily, A. H. (2020). Autonomous micro-robot like sperm based on piezoelectric actuator, *International Conference on Electrical, Communication, and Computer Engineering (ICECCE)*, pp. 1–6.

Rosenberger, L. J. (2001). Pectoral fin locomotion in batoid fishes: undulation versus oscillation, *The Journal of Experimental Biology*, 204, pp. 379–394.

Rosenberger, L. J. and Westneat, M. W. (1999). Functional morphology of undulatory pectoral fin locomotion in the stingray taeniura lymma (Chondrichthyes: dasyatidae), *Journal of Experimental Biology*, 202, pp. 3523–3539.

Sfakiotakis, M., Lane, D. M. and Davies, J. B. C. (1999). Review of fish swimming modes for aquatic locomotion, *IEEE Journal of Oceanic Engineering*, 24(2), pp. 237–252.

Shoele, K. and Zhu, Q. (2010). Numerical simulation of a pectoral fin during labriform swimming, *J. Exp. Biol.*, 213, pp. 2038–2047.

Sitorus, P., Nazaruddin, E. Y., Leksono, E. and Budiyono, A. (2009). Design and implementation of paired pectoral fins locomotion of labriform fish applied to a fish robot, *Science direct, Journal of Bionic Engineering*, 6, pp. 37–45.

Videler, J. J. (1993). Interactions between fish and water. In: *Fish Swimming*, Springer, Dordrecht.

Vogel, S. (1994). Life in moving fluids, Princeton: Princeton University Press.

Walker, J. A. and Westneat, M. W. (1997). Labriform propulsion in fishes: kinematics of flapping aquatic flight in the bird wrasse Gomphosus varius (Labridae), *J. Exp. Biol.*, 200, pp. 1549–1569.

Wang, W., Li, Y., Zhang, X., Wang, C., Chen, S. and Xie, G. (2016). Speed evaluation of a freely swimming robotic fish with an artificial lateral line, *IEEE International Conference on Robotics and Automation (ICRA)*.

Williams, T. (1994). A model of rowing propulsion and the ontogeny of locomotion in artemia larvae, *Biol. Bull.*, 187, pp. 164–173.

Chapter 3

SWIMMING ROBOT DESIGN

3.1. Introduction

As mentioned in Chapter 2, the shape, size, and biomechanical aspects of the pectoral fins have a huge impact on the performance of the swimming in Labriform swimming mode. In addition, the shape, size, and weight of the swimming robot body affect the swimming motion. Recently, one of the major fields of interest of robotics researchers is how to resolve some of the difficulties associated with the different water environments like ponds, lagoons, rivers, and deep waters or even shallow. Sometimes, exploration and searching in many of these environments can be dangerous or inaccessible to human beings. Thus, it is important to develop an efficient technology with direct applications as a solution to the problems mentioned such as the robust design of swimming robots to overcome these challenges. Hence, a large number of studies over the few past years on swimming robots of different sizes and shapes have been an attractive subject for many researchers. A shape that will largely help the pectoral fins in producing a large amount of thrust force when the fins start pushing toward the back of the body while minimizing the drag force to a minimum when the fins return to their original position, is of much interest.

In this chapter, the design, and implementation of a swimming robot propelled by two pectoral fins are achieved. The complete design of the swimming robot body is presented. Next, the effect of varying fin oscillation speed in two phases will be observed, the first phase is the power stroke one, where the fins start to push toward the back of the body, and the recovery phase, where the fins return toward the frontal part of the body. The detailed steps of the robot design are presented and the thrust force exerted by pectoral fins has been evaluated. The suggested swimming robot is a rigid-body with an elliptical cross-sectional area, which helps in minimizing the water resistance during the thrusting process.

For the robust design of the swimming robot, the kinematic and dynamic model should be derived, which is achieved by Newton–Euler equations, further, the hydrodynamic forces that are exerted on the swimming robot have been assessed by using the method of computational fluid dynamics (CFD). To improve the performance of swimming robot motion, a PID controller has been used for this purpose. The validation of the swimming robot design is achieved by simulating the swimming robot design in the SOLIDWORKS® platform, while several practical experiments will be performed for the physical test.

3.2. Construction of the Swimming Robot

This section describes a 3D design of a swimming robot. The hydrodynamic behavior, weight of the robot, biomimetic look, and waterproofness methods are the main issues that are taken into consideration with details. The robot's body consists of two parts: the main body that contains the microcontroller and other electronic units, and the robot's head, which contains a pair of waterproof servomotors, each one is linked to a concave pectoral fin via a joint. All the robot's components including body, joints, and fins are designed by SOLIDWORKS® software and then printed by a 3D printer. The robot body is designed with a cross-sectional area as an elliptical-form where its size gradually decreases through the longitudinal axis of the body to reduce water resistance.

3.2.1. *The Waterproof Body Part*

The body of the robot is designed with a hard shell of thickness 6 mm of Polylactic Acid (PLA). It is designed to be symmetric about the longitudinal axis of the robot body. The main part of the body consists of two parts, the housing part, and the lid part. Utilizing the lip and groove property of plastics in SOLIDWORKS®, the body's lip of 2 mm thickness is built into the mating edge of the housing part, and the groove of 4 mm thickness is built into the edge of the lid as shown in Figure 3.1(a). In this way, when pushing the two parts together, with the use of O-rings, the design will be sealed, waterproofed, and tightly closed.

3.2.2. *The Head Part*

To protect the internal electronics of the robot's body against the water, the servomotors are separated from the rest of the body, where a thin plastic plate is designed and built-in at the inner base of the head. The servomotors are fixed parallel to the longitudinal axis of the body. Fins are attached to the actuators through small specially-designed joints of 1.5 cm in length, and openings are made on both sides of the head to facilitate the movement of the fins in the forward and backward directions during the power and recovery strokes, respectively. To provide double protection of the internal parts of the robot, the inside of the robot's body is coated with high-quality glue as shown in Figure 3.1(b). Furthermore, for extra protection against water, all inner electronic units are covered with NANO PROTECH coating technology spray. The final design is given in Figure 3.1(c).

It is confirmed that tolerance to water conditions exists such as pressure during the power and recovery periods, where a simple test is done at the maximum speed of the servomotors under pressure 101325 Pa. It is noted that the maximum amount of pressure rises by a very small percentage at the frontal part of the body in the case of power stroke as shown in Figure 3.2(a), when the movement of the robot is against the direction of the water flow, it reaches 101325.04 Pa. This difference of 0.04 is considered very little and can be neglected. Likewise, for the case concerning movement during the recovery stroke when the body moves in the opposite direction to that in the power stroke, so the highest amount of pressure at the back part of the body is of the same value as in power stroke motion as shown in Figure 3.2(b).

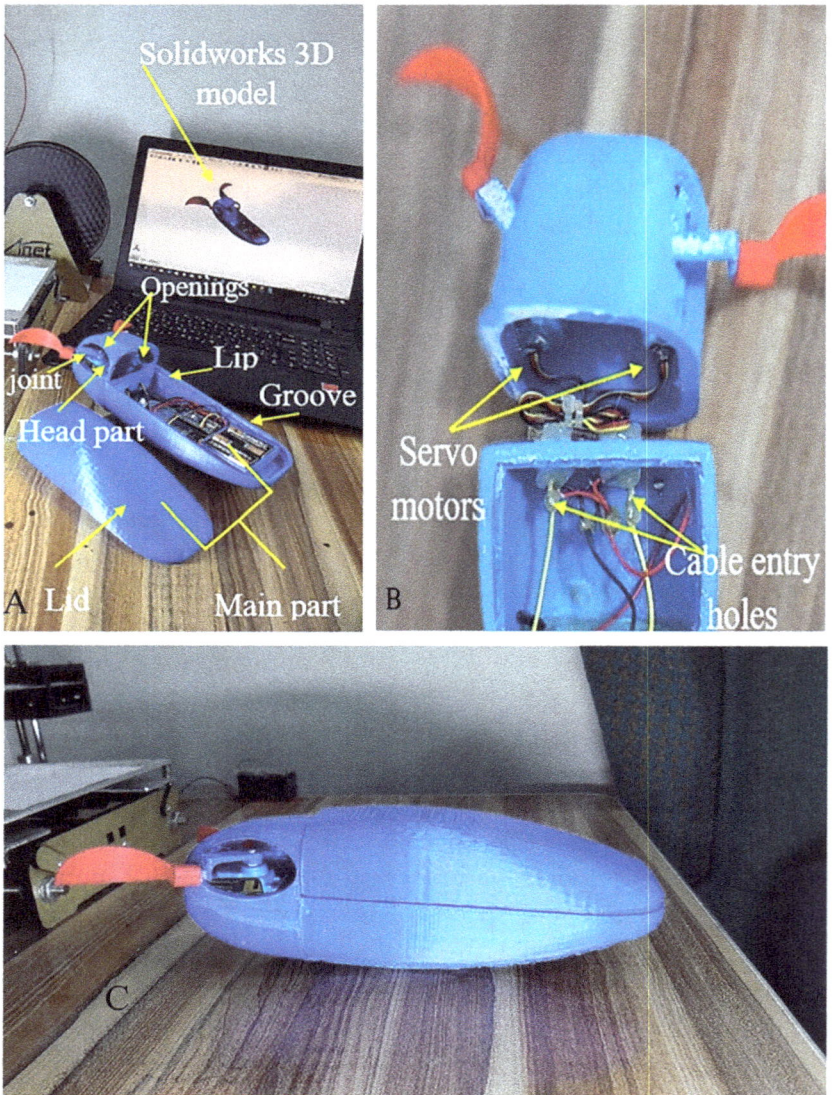

Figure 3.1. Proposed design of swimming robot. (a) The inner design of the robot, (b) The two parts of the robots, (c) The complete 3D robot prototype.

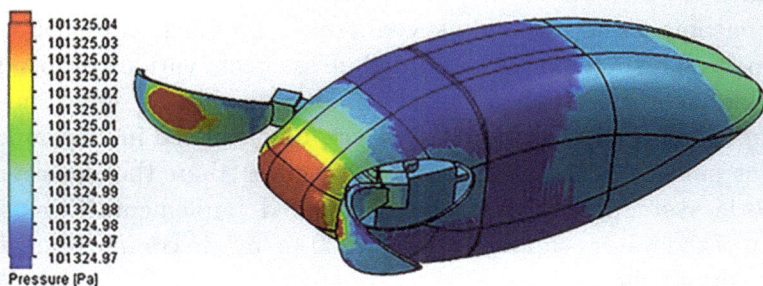

(a)

(b)

Figure 3.2. Pressure distribution on the robot's surface. (a) Power stroke. (b) Recovery stroke.

3.3. Hydrodynamic Reactive Force and Moment Evaluation

The main hydrodynamic impacts on an object during its movement in water will be summarized. The hypothesis of fluid hydrodynamics is somewhat complex and it is hard to implement a confident model for all of the hydrodynamic effects. The present work focuses mainly on the effect of added mass only, while other effects are out of the scope of this study.

When a body moves in water, there exists extra inertia of the water encompassing the body speeding up as a reaction due to this movement. This reaction is the added mass effect. Various properties hold relative to the (6×6) of the inertia matrix in describing a rigid body. This effect of added mass can be dismissed in mechanical robotics since the air density is a lot lighter than the density of a robotic system. While, in the submerged implementations, the density of the water is approximately $1000 \, \text{kg/m}^3$, in comparison with the density of the vehicles.

The added mass definition has the same impact as the dynamical system with an extra mass, in which the value of the generated force due to the added mass is proportional to the acceleration multiplied by a factor that depends on the volume of the body and its geometrical shape. The proposed design of the robot's pectoral fin is concave, a one-eighth of a hollow spherical shape (Fin2 in the previous chapter). This design will allow a portion of water to move along within the body according to the fin movement. Various properties affect the movement of water in a real domain. In order to derive a mathematical model and comprehend the behavior of water flow, those properties need to be precisely examined to provide the exact transition between the numerical calculation domain and reality. Pressure, temperature, velocity, viscosity, and density are the main properties that affect the flow of water and should be examined simultaneously. By solving Navier–Stokes equations adopted by SOLIDWORKS®, it is applicable to describe the change in the water properties.

3.4. Rigid Body Dynamics

Two frames need to be defined which are the body's fixed reference frame $F_B = [x_b y_b z_b]^T$, in which the origin coincides with the center of buoyancy, and the earth reference frame $F_E = [X_E Y_E Z_E]^T$, where T refers to the transpose of the matrix as shown in Figure 3.3. Consider the following vectors: $\xi = [\xi_1^T \xi_2^T]^T$, $\xi_1 = [x, y, z]^T$, $\xi_2 = [\phi, \theta, \Psi]^T$ where ξ represents the position and orientation coordinates with respect to the earth-fixed reference frame F_E. The

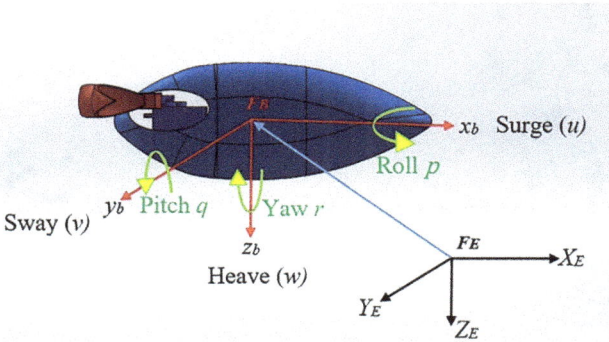

Figure 3.3. Body and earth reference frames.

velocity vector is defined as V with linear V_{b1} and angular V_{b2} velocities, whereas the force/momentum vectors are defined as F_b and M_b, respectively. These vectors can be given as follows:

$$V = [V_{b1}, V_{b2}]^T, \quad V_{b1} = [u, v, w_z]^T, \quad V_{b2} = [p, q, r]^T$$

$$F_b = [F_{bx}, F_{by}, F_{bz}]^T, \quad M_b = [M_{bx}, M_{by}, M_{bz}]^T, \quad \Gamma = [F_b, M_b]^T$$

where Γ represents force/moments in vector form. V_{b1} and V_{b2} are the linear velocity components and angular velocity components, such that $u, v,$ *and* w_z are the surge velocity, sway velocity and heave velocity, respectively. While $p, q,$ and r describe the roll angular velocity, pitch angular velocity and yaw angular velocity, respectively. The robot's body motion with respect to the earth-fixed frame is given by a velocity transformation:

$$\dot{\xi}_1 = Q_1(\xi_2)V_{b1} \tag{3.1}$$

where $Q_1(\xi_2)$ is a transformation matrix that relates to Euler angles roll ϕ, pitch θ, and yaw Ψ. In order to find $Q_1(\xi_2)$, it is described by three rotations as described below (see Figure 3.4). Consider the following coordinates:

$$F_{B1} = [x_1, y_1, z_1]^T, \quad F_{B2} = [x_2, y_2, z_2]^T \quad \text{and} \quad F_{B3} = [x_3, y_3, z_3]^T$$

The coordinate F_{B3} is obtained by translating the earth-fixed coordinate system $F_E = [X_E \, Y_E \, Z_E]^T$ parallel to itself until its origin

(a)   (b)

(c)

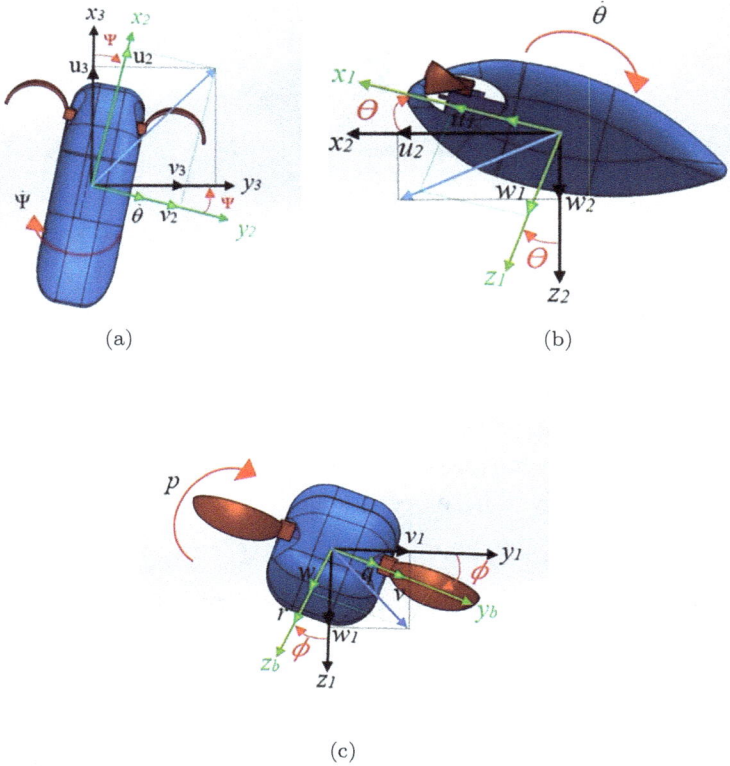

Figure 3.4. xyz rotational sequence showing the components of both linear velocity ($u, v,$ and w) and angular velocity ($p, q,$ and r). (a) Head rotation over angle Ψ about z_3-axis where $w_3 = w_2$. (b) Pitch rotation over angle θ about y_2-axis where $v_1 = v_2$. (c) Roll rotation over angle ϕ about x_1-axis where $u_1 = u_2$.

coincides with the origin of body reference frame $F_B = [x_b, \ y_b, \ z_b]^T$. The resulting frame $F_{B3} = [x_3, \ y_3, \ z_3]^T$ will be rotated about z_3 by a yaw angle Ψ to produce another frame $F_{B2} = [x_2, \ y_2, \ z_2]^T$, which will in turn be rotated about y_2-axis by a pitch angle θ and yielding a new reference frame $F_{B1} = [x_1, \ y_1, \ z_1]^T$, this time the resulting frame will be rotated by a roll angle ϕ about the x_1-axis, producing the body coordinate reference frame $F_B = [x_b, \ y_b, \ z_b]^T$.

This sequence of linear velocity translations is not arbitrary; instead, it uses xyz-convention in terms of Euler's angles for rotation and can be expressed as:

$$Q_1(\xi_2) = E_{z,\Psi}^T E_{y,\theta}^T E_{x,\phi}^T \tag{3.2}$$

$$E_{x,\phi} = \begin{bmatrix} 1 & 0 & 0 \\ 0 & \cos\phi & -\sin\phi \\ 0 & \sin\phi & \cos\phi \end{bmatrix} \tag{3.3}$$

$$E_{y,\theta} = \begin{bmatrix} \cos\theta & 0 & \sin\theta \\ 0 & 1 & 0 \\ -\sin\theta & 0 & \cos\theta \end{bmatrix} \tag{3.4}$$

$$E_{z,\Psi} = \begin{bmatrix} \cos\Psi & -\sin\Psi & 0 \\ \sin\Psi & \cos\Psi & 0 \\ 0 & 0 & 1 \end{bmatrix} \tag{3.5}$$

The orientation of the robot's body relative to the earth-fixed reverence frame is given as:

$$V_{b2} = \begin{bmatrix} \dot\phi \\ 0 \\ 0 \end{bmatrix} + E_{x,\phi}\begin{bmatrix} 0 \\ \dot\theta \\ 0 \end{bmatrix} + E_{x,\phi}E_{y,\theta}\begin{bmatrix} 0 \\ 0 \\ \dot\Psi \end{bmatrix} \tag{3.6}$$

To study the motion of the robot, the body's dynamics are defined based on Kirchhoff's relations of motion of conservation of linear momentum M_L and an angular momentum M_A in an inviscid fluid as in:

$$\dot{M_L} = M_L \times V_{b2} + F_b \tag{3.7}$$

$$\dot{M_A} = M_A \times V_{b2} + M_L \times V_{b1} + M_b \tag{3.8}$$

where F_b and M_b are the total forces generated and moments exerted by the body at the center of buoyancy, respectively.

$$M_L = MV_{b1} + D^T V_{b2} \tag{3.9}$$

$$M_A = DV_{b1} + IV_{b2} \tag{3.10}$$

where M can be defined as the mass matrix, D refers to Coriolis and centripetal matrix, and I is considered as the inertia matrix. These matrices represent the effect of the added mass due to fluid impact

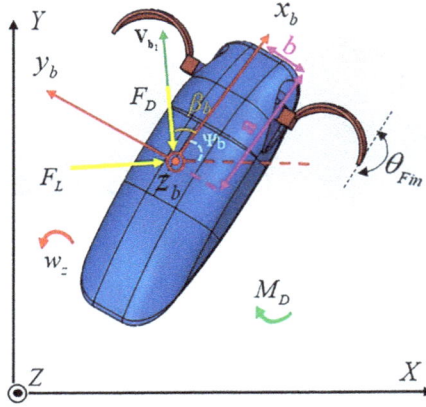

Figure 3.5. The free body diagram of the swimming body.

and the body's inertia. It is assumed that the robot is neutrally buoyant, and both centers of mass and buoyancy are of the same projection on the $x_b y_b$-plane. The focus is on the planer motion only, that is, three degrees of freedom are taken into consideration, which are namely, surge, sway, and yaw as mentioned earlier as in Figure 3.5.

The angle of attack β_b can be defined as the angle that the linear velocity vector V_b, having a magnitude of $|V_{b1}| = \sqrt{(u^2 + v^2)}$ forms with the x_b-axis where $\tan(\beta_b) = v/u$ and the direction of the robot relative to the earth frame is given by the angle Ψ_b. Furthermore, it is assumed that the robot's body is symmetrical about the $x_b z_b$-plane. In this case, the off-diagonal elements of the inertia matrix with respect to y_b-axis will be set to zero, and by ignoring the inertial coupling between surge velocity, sway velocity, and yaw velocity, then equations (3.7) and (3.8) will be reduced to the following:

$$\left. \begin{aligned} (M_b - M_{\dot{u}a})\dot{u} &= (M_b - M_{\dot{v}a})vr + F_{bx} \\ (M_b - M_{\dot{v}a})\dot{v} &= -(M_b - M_{\dot{u}a})ur + F_{by} \\ (I_z - N_r)\dot{r} &= M_{bz} \end{aligned} \right\} \qquad (3.11)$$

where M_b is a robotic fish mass, I_z is the robot inertia about the z_b-axis, $M_{\dot{u}a}$, $M_{\dot{v}a}$ and $N_{\dot{w}z}$ are the effect of added mass (and or) inertia on the rigid body. Specifically, $M_{\dot{u}a}\dot{u}$ represents the hydrodynamic added mass force in the x_b-direction due to an acceleration $\dot{u}$ along that axis; in the same manner, $M_{\dot{v}a}\dot{v}$ and $N_{\dot{w}z}\dot{w}_z$ can be defined. In many practical experiments, $-M_{\dot{u}a}$, $-M_{\dot{v}a}$, and $-N_{\dot{r}}$ are positive numbers. In equation (3.11) F_{bx}, F_{by}, and M_{bz} are the forces in $x_b y_b$, and z_b directions, respectively, and M_{bx}, M_{by}, and M_{bz} are the moments about x_b, y_b, and z_b respectively. The linear and angular momentum can be given as follows:

$$\left.\begin{aligned}
F_{bx} &= F_{bxh} - F_D \cos \beta_b + F_L \sin \beta_b \\
F_{by} &= F_{byh} - F_D \sin \beta_b - F_L \cos \beta_b \\
M_{bz} &= M_{bzh} + M_D
\end{aligned}\right\} \tag{3.12}$$

where F_{bxh}, F_{byh}, and M_{bzh} are the hydrodynamic forces (and or) moments transmitted from the pectoral fins to the robot's body. F_D, F_L and M_D, are the body's drag force, lift force and moment force, respectively. These forces are given as:

$$\left.\begin{aligned}
F_D &= 1/2\rho V_{b1}^2 S C_D \\
F_L &= 1/2\rho V_{b1}^2 S C_L \\
M_D &= -C_M r^2 \mathrm{sgn}(r)
\end{aligned}\right\} \tag{3.13}$$

where ρ is the water density, S is the projected surface area. C_D, C_L and C_M are the drag, lift and moment coefficients, respectively. sgn(.) is the signum function. The kinematics of the robot is given as:

$$\left.\begin{aligned}
\dot{X} &= u \cos \Psi_b - v \sin \Psi_b \\
\dot{Y} &= v \cos \Psi_b + u \sin \Psi_b \\
\dot{\Psi}_b &= r
\end{aligned}\right\} \tag{3.14}$$

where Ψ_b is the angle that lies between the x_b-axis of the body and X-axis of global reference frame.

Generally, the drag force is the force that is in the opposite direction to the flow and the lift force is the normal force to the flow. In order to calculate the hydrodynamic force applied by the water, the proposed model is set to be a stationary part and let the water

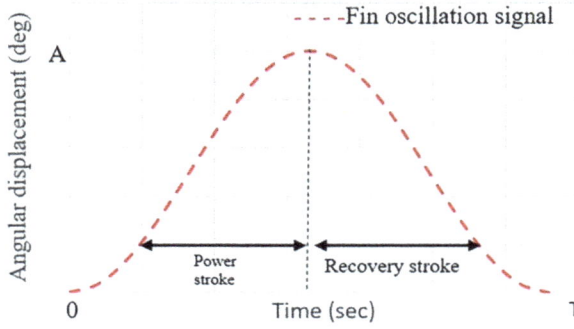

Figure 3.6. Servomotor rotation angle.

be the moving part. The servomotor oscillates back and forth by θ_{Fin} as shown in Figure 3.6 based on equation (2.13).

The added masses, inertia and body's wetted surface area were calculated by considering the robot's body as a prolate spheroid accelerating in the water. The body inertia is evaluated about z_b-axis as:

$$I_z = \frac{1}{5} M_b (a^2 + b^2) \tag{3.15}$$

$$a = \frac{\text{Length of the body}}{2} \tag{3.16}$$

$$b = \frac{\text{Width of the body}}{2} \tag{3.17}$$

where a and b are the lengths of the semi defined axis of the swimming body. The effect of added masses and effect of added inertia can be found as follows:

$$M_{\dot{u}a} = -K_1 M_b \tag{3.18}$$

$$M_{\dot{v}a} = -K_2 M_b \tag{3.19}$$

$$N_{\dot{w}z} = -K_3 I_z \tag{3.20}$$

where the positive constants $K_1, K_2,$ and K_3 are Lamb's K-factors that depend only on the geometrical shape of the body and can be defined as follows:

$$K_1 = \frac{\alpha^*}{2 - \alpha^*} \tag{3.21}$$

$$K_2 = \frac{\beta^*}{2 - \beta^*} \tag{3.22}$$

$$K_3 = \frac{e^4(\beta^* - \alpha^*)}{(2 - e^2)[2e^2 - (2 - e^2)(\beta^* - \alpha^*)]} \tag{3.23}$$

where α^*, β^*, and e can be expressed as follows:

$$\alpha^* = \frac{2(1 - e^2)}{e^3}\left[\frac{1}{2}\ln\left(\frac{1 + e}{1 - e}\right) - e\right] \tag{3.24}$$

$$\beta^* = \frac{1}{e^2} - \frac{1 - e^2}{2e^3}\ln\left(\frac{1 + e}{1 - e}\right) \tag{3.25}$$

$$e = \sqrt{1 - \left(\frac{b}{a}\right)^2} \tag{3.26}$$

3.5. Improvement of the Pectoral Fin Performance

A 3D-model of the robot is designed by SOLIDWORKS®, and then exported to MATLAB/Simulink via the Simscape add-in tool. Figure 3.7 shows the Simscape construction of the proposed model.

The complete design model is simulated by MATLAB in order to observe the response of fins' performance. The signal is fed to the system as a math function to obtain the pattern given in Figure 3.8, where the starting angle is set at 0.872 rad which is a translation of 50°. The output of the proposed system is measured by a scope tool to obtain the position and velocity.

The responses of both positions in Figure 3.9(a) and velocity in Figure 3.9(b) show a noticeable ripple that may affect the overall performance of the system performance, so, the Proportional-Integral-Derivative controller (PID) has been employed as a controller to improve the performance of the system. PID controller can be

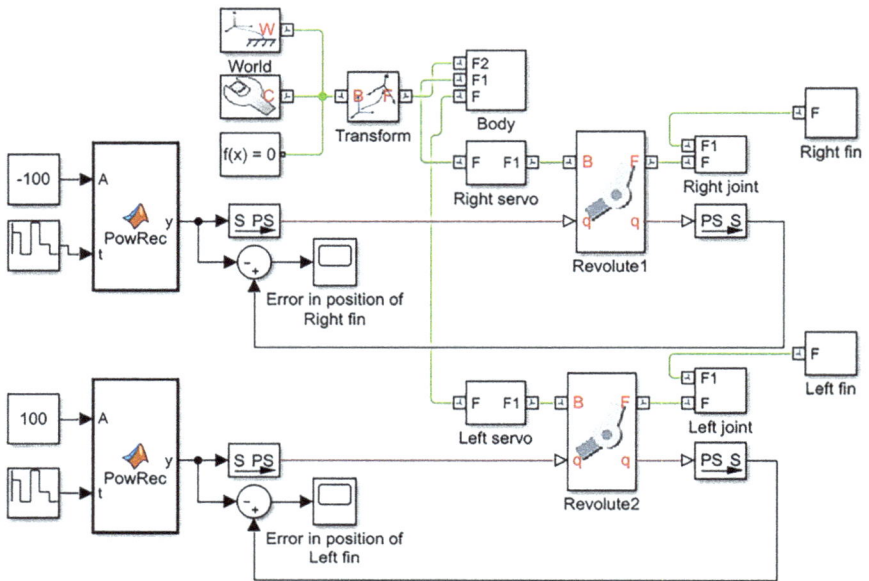

Figure 3.7. Simscape model without PID controller.

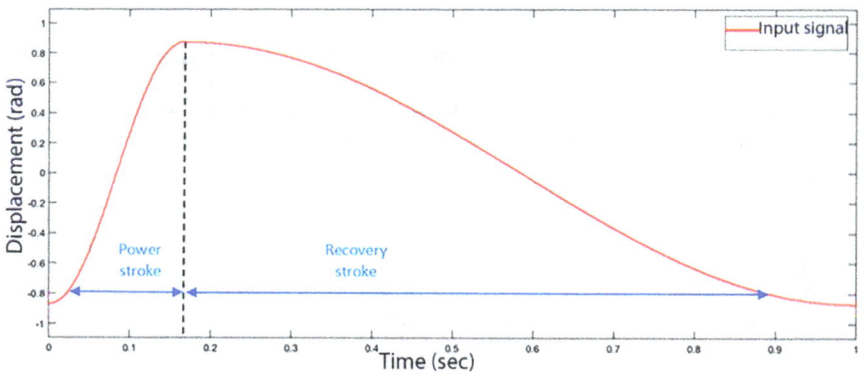

Figure 3.8. Input signal.

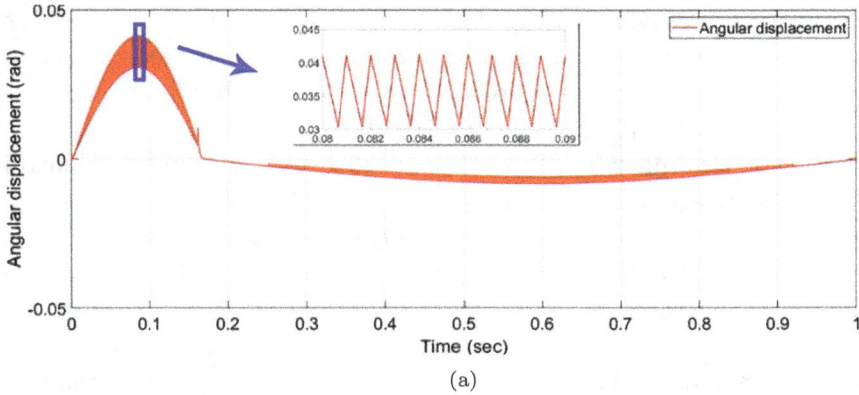

(a)

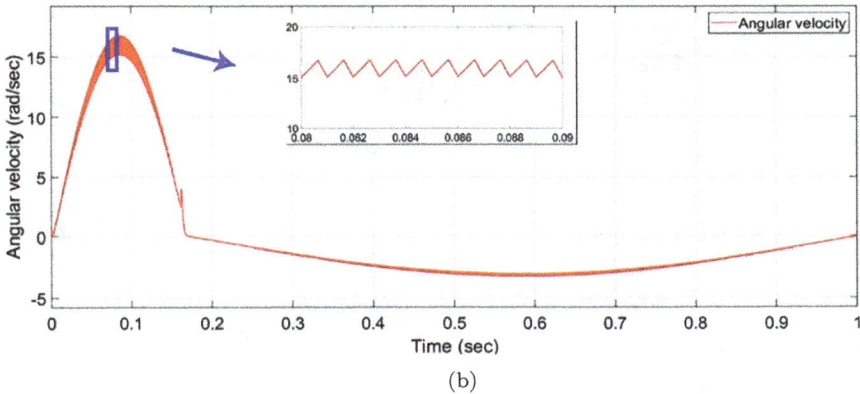

(b)

Figure 3.9. System responses without PID controller. (a) Error in position signal. (b) Output velocity signal.

expressed as:

$$E(t) = K_P e_r(t) + K_I \int_0^t e_r(t)dt + K_D \frac{de_r(t)}{dt} \qquad (3.27)$$

where $E(t)$ is the output signal of the PID controller, $e_r(t)$ is given as the difference between the desired value and the actual one, K_P, K_I, and K_D are the proportional, integral and derivative gains, respectively.

The proportional part is used to improve the open loop gain and to reduce the steady-state error of the overall system. The integration part can enhance the type and strengthen the robustness of the system, whereas derivation part is to improve the stability of the system and produce correction signal early effective to increase the degree of damping.

The PID tuner application provided by Simulink® Control Design software, automatically tunes the gains of a PID controller for a single input single output (SISO) system to achieve a balance between performance and robustness. Since the revolute joints are identical, the right fin is analyzed only, and the same results can be obtained for the left one. The same input signal given earlier in Figure 3.8 is used. The complete system along with the controller is shown in Figure 3.10.

There are large improvements in the system performance in terms of position and velocity. It can be noticed that it could reach about zero after 0.18 sec as shown in Figure 3.11. The PID controller gains are tuned so that $K_P = 0.516533$, $K_I = 203.985172$, and

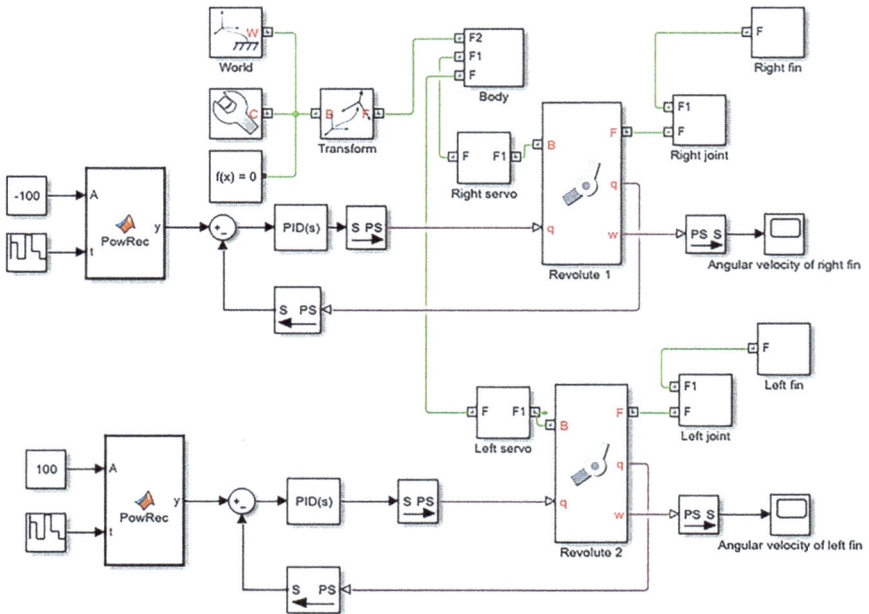

Figure 3.10. Simscape model with PID controller.

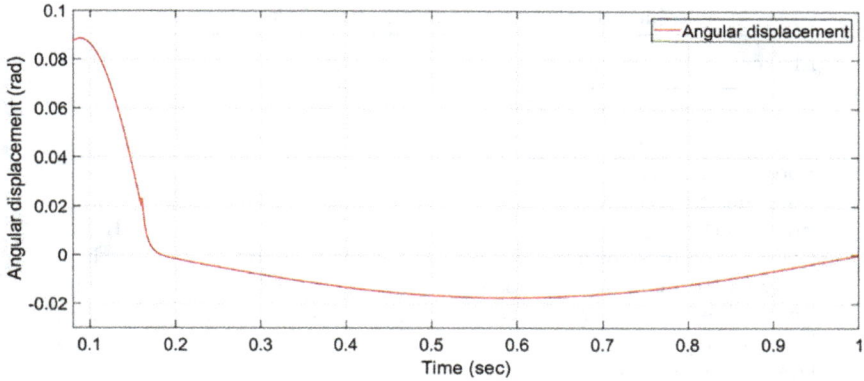

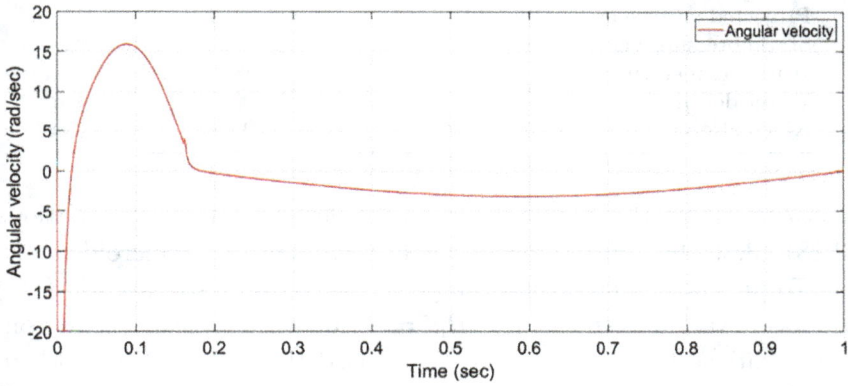

Figure 3.11. System responses with PID controller. (a) Error in position signal. (b) Output velocity signal.

$K_D = -0.00045557$. The servomotor maximum limit is translated at a maximum actuation frequency of 1.515 Hz that correspond to 0.66 sec, the simulation time is extended to 1 sec in order to be able to capture the performance trend of the whole system.

3.6. Parameter Identification

The parameters used throughout this chapter is derived either directly or empirically and are listed as shown in Table 3.1. The added

Table 3.1. Parameter identification of the robot.

Element	Measured value	Unit
Body Length (L)	0.185	m
Body width (W)	0.06	m
Body height (H)	0.065	m
Body mass (M_b)	0.650	Kg
Body inertia (I_z)	12.29×10^{-4}	Kg/m^2
Body added mass over x_b-axis ($M_{\dot{u}a}$)	0.07635	Kg
Body added mass over y_b-axis ($M_{\dot{v}a}$)	0.52574	Kg
Body added inertia around z_b-axis ($N_{\dot{w}z}$)	5.916×10^{-4}	Kg/m^2
Fin length (PF_L)	0.05	m
Drag force coefficient (C_D)	0.4	—
Lift force coefficient (C_L)	3.4	—
Momentum force coefficient (C_M)	1.52×10^{-3}	Kg/m^2
Projected surface area of the robot (S)	0.0764	m^2
Static pressure via CFD	101325	Pa
Water temperature	293.2	K
Water density	1000	Kg/m^3
Kinematic viscosity of water	10^{-6}	m^2/sec

mass/inertia of the swimming robot is derived following equations (3.18)–(3.26) by approximating the body to a prolate spheroid shape.

To produce a net thrust, the power to recovery stroke ratio (R_F) has been varied respectively, $R_F = 1$ for symmetrical oscillation. Other ratios, $R_F = 2, 3, 4$, and 5 are used throughout this work to identify the optimum one in thrust generating. For $R_F = 1$, the optimum starting angle is defined with the aid of CFD calculations. The maximum frequency translated from the servomotor specifications is used to the starting angle of oscillation at 1.5151 Hz. Equations (2.9)–(2.11) are adopted for the calculation of the hydrodynamic forces and coefficients.

3.7. Simulation and Experimental Results

To evaluate the swimming robot's compliance with the performance objectives, two tests have been carried out throughout this work. The first test aims to examine the robot's ability to swim in a straight-line path. This was implemented by placing the robot in a swimming pool of $(1.00 \times 0.65 \times 0.65)$ meter dimensions of (length, width, and

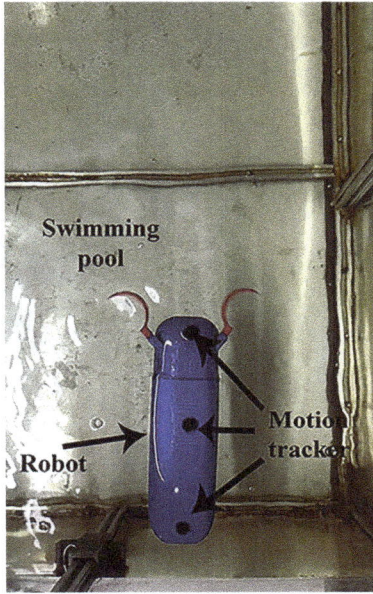

Figure 3.12. Swimming robot in the pool (Top view).

height), respectively. A top view of the robot in the swimming pool is shown in Figure 3.12. Evaluation of robot's motion is achieved by placing a camera for the top view of the robot at a distance of 0.5 m.

On the other hand, the second test was designed to determine if the model could swim at variable speeds controlled by the user. Based on Computational Fluid Dynamics (CFD) offered by flow simulation from SOLIDWORKS®, it is able to calculate some very attractive parameters. To add a controlling mechanism and making use of the proposed PID controller method, a simple modification is done on the servomotor control circuit as shown in Figure 3.13.

3.7.1. *The Effect of the Starting Angle*

The effect of fin beat is examined as shown in Figure 3.14. The total drag force exerted by varying fin angles is presented in Figure 3.15. The angle is varied from $\theta'_{\text{Fin}} = 0°$ to $\theta'_{\text{Fin}} = 50°$ at a step of $10°$. The obtained drag decreases linearly, where the highest drag is recorded when the fin is perpendicular to the water flow direction at $\theta'_{\text{Fin}} = 0°$.

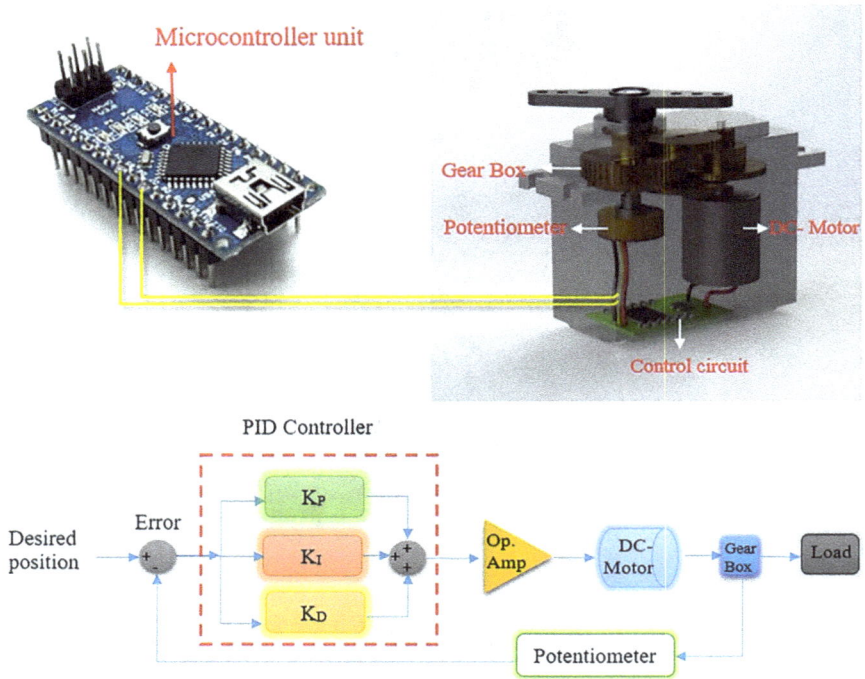

Figure 3.13. Servomotor modification.

In this manner, the best angle of pectoral fin beat at power stroke is set to be 50°. Due to the design specifications, this angle is considered as the maximum value in producing the highest thrust. All angle tests were carried out with maximum servomotor specifications of 1.515 Hz of frequency and 0.11°/60 sec of speed.

3.7.2. *Power to Recovery Ratio Effect*

In rowing motion of Labriform mode, during the power stroke, the thrust force should be at a maximum value, while in recovery stroke, the drag, should be kept to a minimum. The proposed model was tested for different power to recovery stroke ratios (R_F) (i.e. a ratio of 1:1, 2:1, 3:1, 4:1, and 5:1).

Experimental analysis for the forward velocity of the robot's body is presented here. Figure 3.16 shows the results when the power

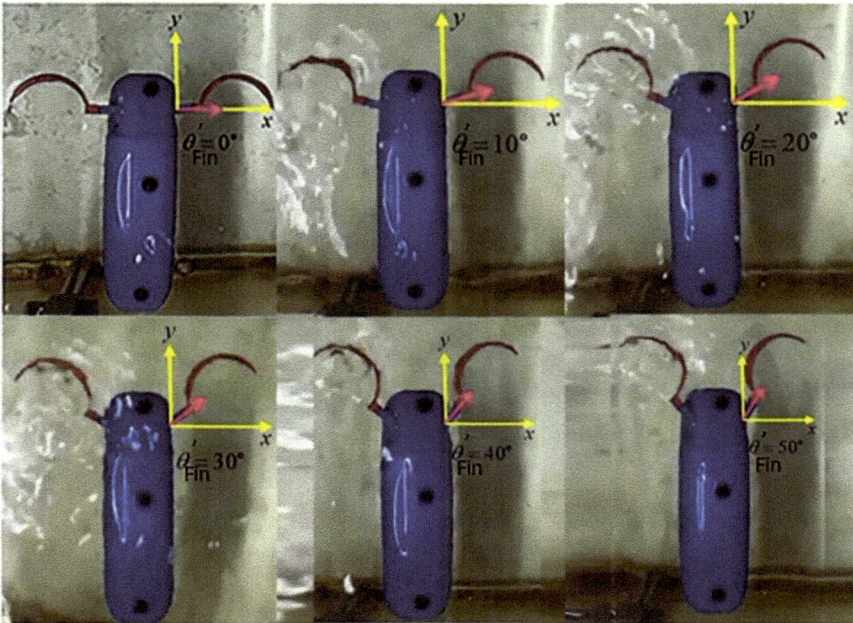

Figure 3.14. Pectoral fin with different starting angles.

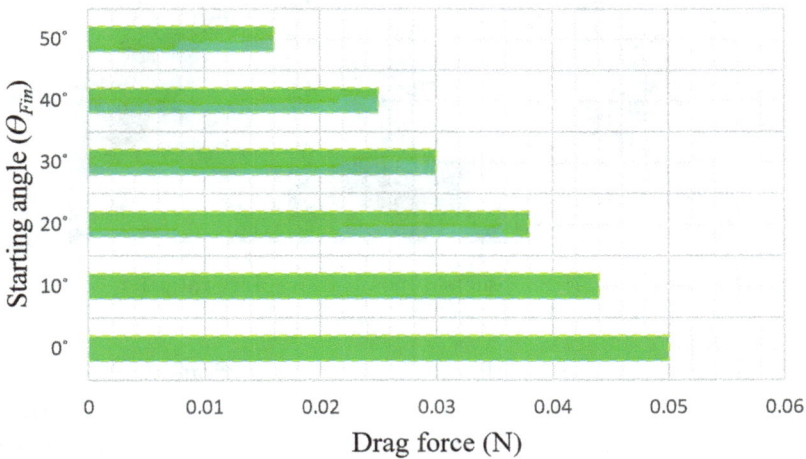

Figure 3.15. Drag force exerted by pectoral fin at different starting angles.

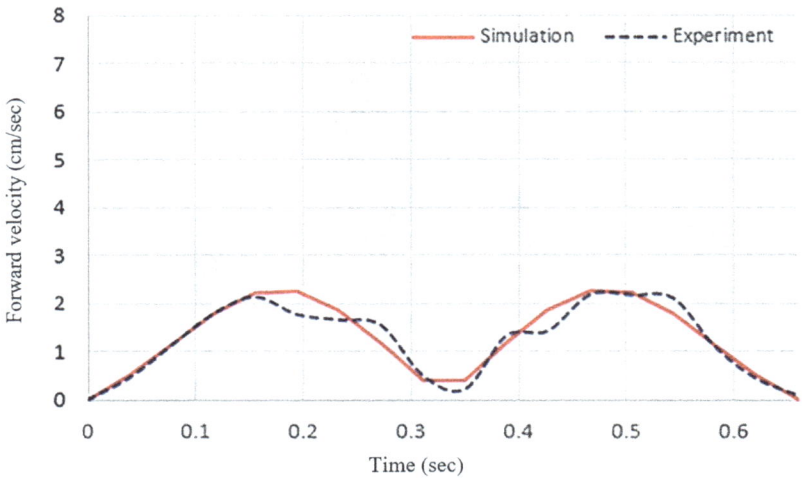

Figure 3.16. Forward velocity at power to recovery ratio 1:1.

Figure 3.17. Snapshots of power to recovery ratio 1:1.

to recovery ratio R_F is 1:1, as mentioned previously, the effect of the hydrodynamic drag force was approximately equal to the thrust force, so the forward velocity at 1:1 would be very small and there is no noticeable velocity gain. A snapshot of 1:1 is given in Figure 3.17, the power stroke cycle is completed during the first 0.33 sec, while recovery stroke will start from 0.33 sec to 0.66 sec.

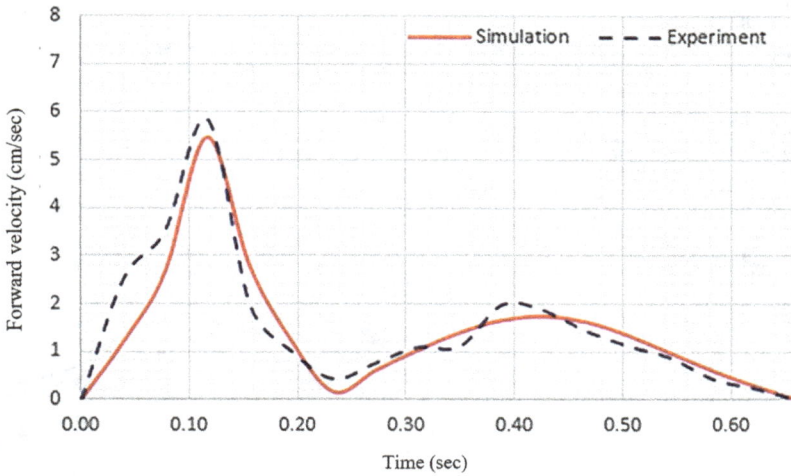

Figure 3.18. Forward velocity at power to recovery ratio 2:1.

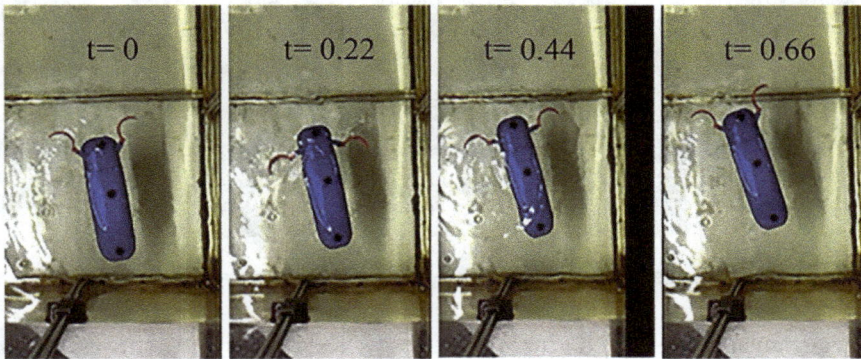

Figure 3.19. Snapshots of power to recovery ratio 2:1.

In contrast, at power to recovery ratio of 2:1 ratio, a noticeably large difference between the thrust and drag forces is seen with results in forward velocity of about 3.8 cm/sec at power and recovery stroke time as in Figure 3.18. A snapshot is given in Figure 3.19, during the first 0.22 sec, the power stroke is over and a recovery stroke starts, the gained forward velocity is interesting as expected, the experimental results are higher than free-swimming simulation results, due to the

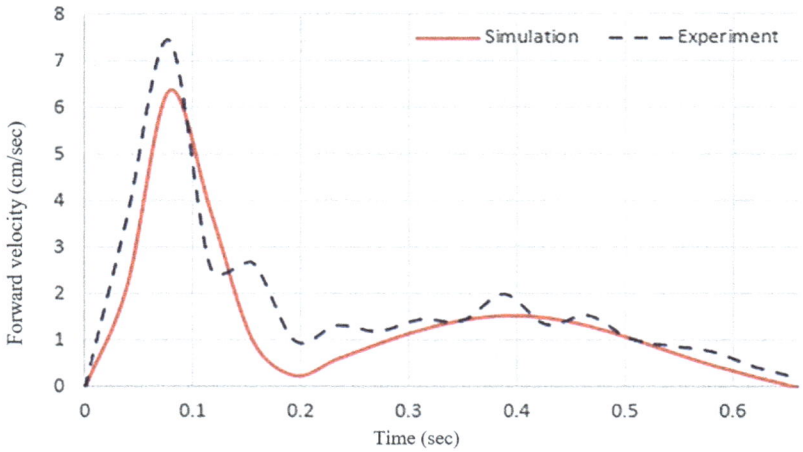

Figure 3.20. Forward velocity at power to recovery ratio 3:1.

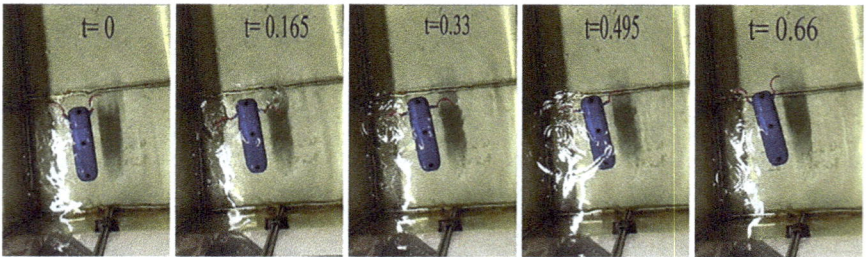

Figure 3.21. Snapshots of power to recovery ratio 3:1.

large thrust generated during the power stroke cycle, the robot is able to swim for few centimeters.

Speeding up the ratio to 3:1 will result in almost the highest magnitude of forwarding velocity as shown in Figure 3.20. Snapshots of Figure 3.21 demonstrate the swimming phases, where the power stroke started from 0 sec to 0.165 sec, the highest speed reached about 5.4 cm/sec, while the recovery stroke will continue from 0.165 sec to 0.66 sec with minimum drag force.

Also for ratio 4:1, an accepted value of swimming velocity is produced as in Figure 3.22. Figure 3.23 shows the snapshots of 4:1 ratio, where the forward velocity is about 4.2 cm/sec. There is a

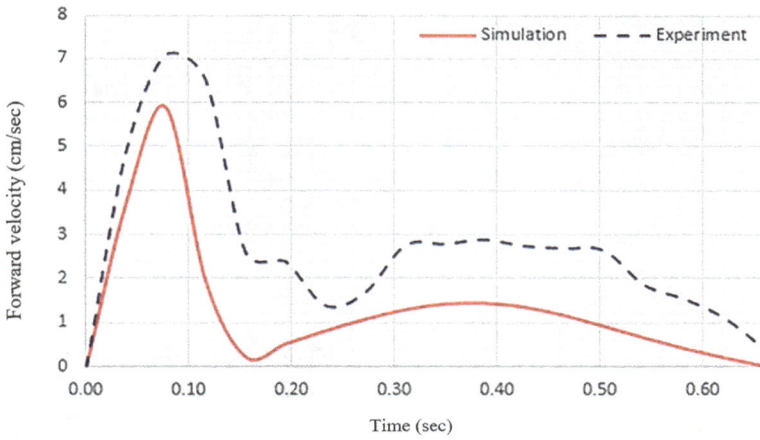

Figure 3.22. Forward velocity at power to recovery ratio 4:1.

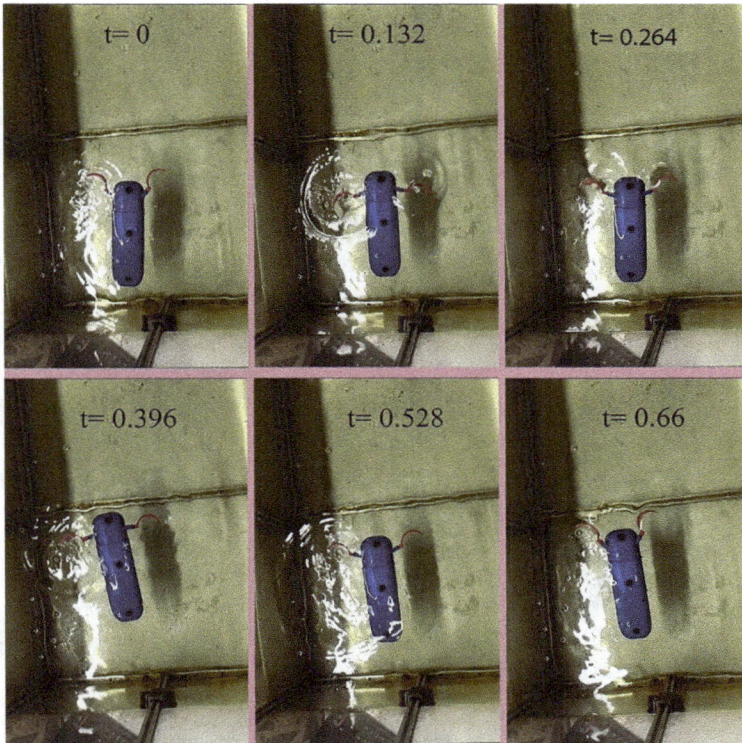

Figure 3.23. Snapshots of power to recovery ratio 4:1.

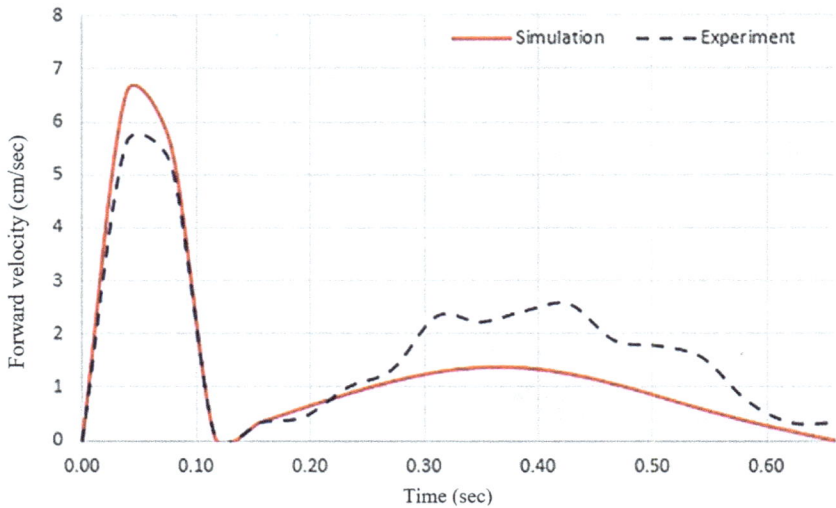

Figure 3.24. Forward velocity at power to recovery ratio 5:1.

noticeable degradation in the velocity at ratio 5:1, as in Figures 3.24 and 3.25, the power phase will start from 0 sec and end with 0.11 sec, producing a forward velocity of about 5.8 cm/sec at power stroke and 2.8 cm/sec at recovery stroke.

Although the simulation results show a positive correlation relationship between the ratio of power to recovery speed and forward velocity of the body, but the calculation of drag force have shown that the drag force at power stroke time increases for the ratio of 1:1, 2:1, 4:1 and 5:1, while it is at minimum in the moderate ratio of 3:1 as in Figure 3.26. Finally, to further clarify the effectiveness of the proposed concave-shape of the pectoral fins, a compilation of the swimming speeds in some focused research is reported in Table 3.2 where (BL/sec) stands for body length per sec.

3.7.3. *Frequency Effect*

Five different simulation conditions have been performed at different fin oscillation frequencies, where different ratios of power to recovery

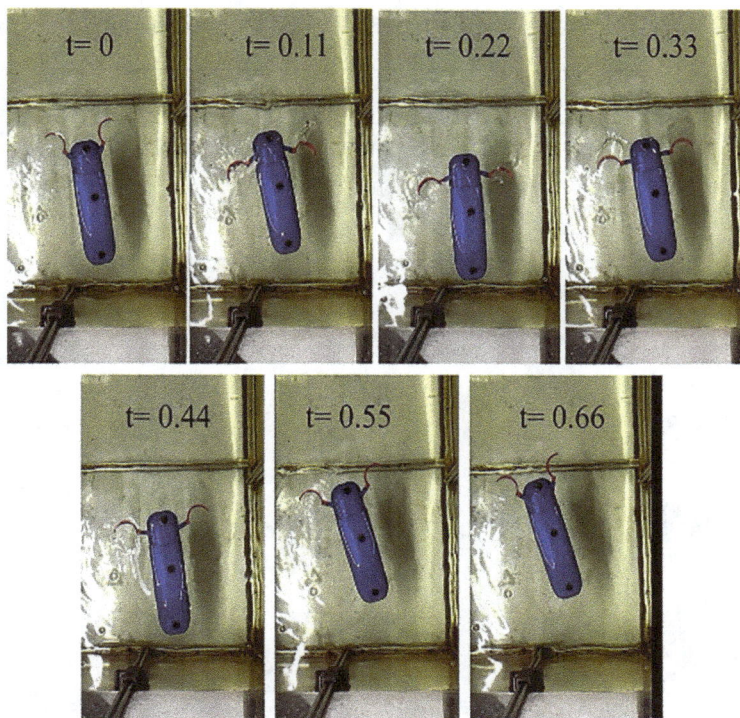

Figure 3.25. Snapshots of power to recovery ratio 5:1.

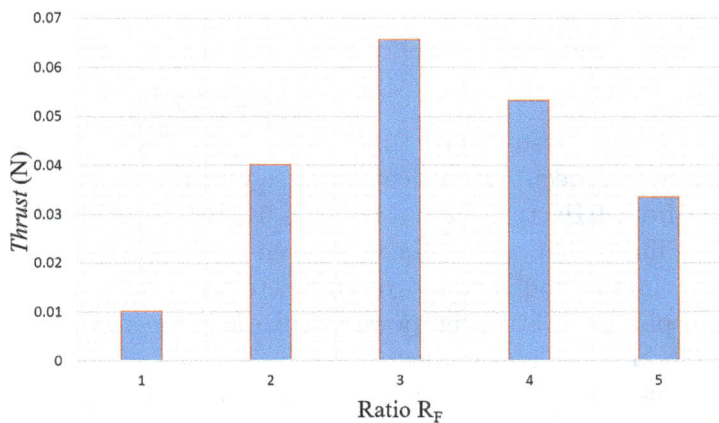

Figure 3.26. Hydrodynamic thrust generated at different R_F.

Table 3.2. The comparison of swimming performance with other works.

References	Forward swimming speed
Ref. [35]	0.045 m/sec (0.33 BL/sec)
Ref. [36]	0.040 m/sec (0.27 BL/sec)
Ref. [89]	0.045 m/sec (0.30 BL/sec)
Proposed System	0.074 m/sec (0.40 BL/sec)

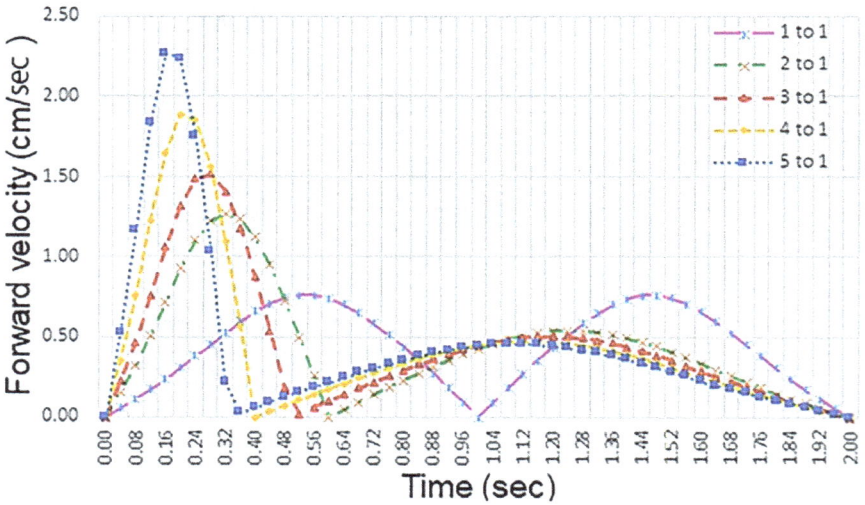

Figure 3.27. Forward velocity of free swimming robot at frequency 0.5 Hz.

stroke (R_F) are tested with different fin oscillation frequencies.
Specifically, the pectoral fin beat frequency of the swimming robot
is varied from 0 Hz to 0.5 Hz, 1 Hz, and 1.5 Hz as in Figures 3.27–
3.29. The fin oscillation frequency is given as $f = 1/T_c$, where T_c is
the total time period required to complete one cycle. To show the
complete phases of both the power stroke and the recovery stroke,
the simulation time is extended to 2 sec.

Utilizing motion manager package, provided by SOLID-
WORKS®, the required motor torque is calculated at the highest

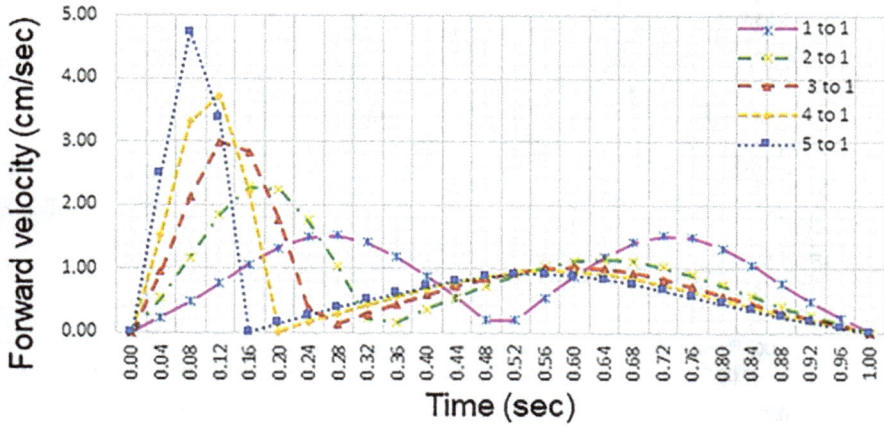

Figure 3.28. Forward velocity of free swimming robot at frequency 1 Hz.

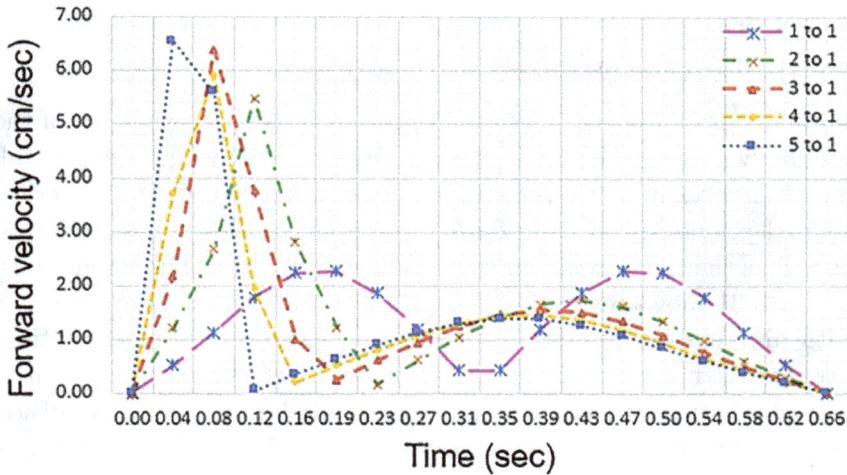

Figure 3.29. Forward velocity of free swimming robot at frequency 1.5 Hz.

speed of $R_F = 3$ and a maximum frequency of 1.515 Hz. The results show a value of 0.228 N.m as in Figure 3.30. This value can be translated to match approximately 3 Kg-cm, where a Hitec 35086W HS-5086WP waterproof digital servomotor is proposed to be used.

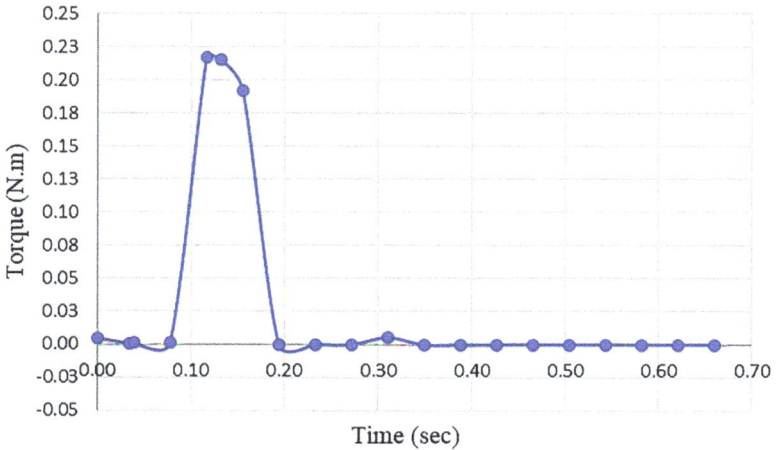

Figure 3.30. Servomotor torque calculation.

3.8. Performance in Terms of Dimensionless Numbers

3.8.1. *Strouhal Number*

The dimensionless numbers are important in the investigation of the similarity between the biomechanics and physical systems, despite all the differences in motion medium or their scale. Strouhal number, as defined previously in Chapter 2, is a dimensionless parameter that describes the tail or wing kinematics of swimming and flying animals. Figure 3.31 shows the optimal range of Strouhal numbers of the robot with different fin oscillation frequencies at the steady state motion. Note that this range is higher than the real fish range, which is typically between 0.25 and 0.5. The reason behind it is the propulsion mechanism that is used in the design of the swimming robot depends on pectoral fins only, so the total forward swimming velocity is relatively low, which results in higher values of Strouhal number. The obtained results are compared with those in the literature at (1 Hz) and the designed system showed a relatively lowest value of Strouhal number in comparison with the others as seen in Table 3.3.

3.8.2. *Reynolds Number*

The Reynolds number is used to determine whether the fluid is a laminar or turbulent flow. If the Reynolds number is less than or

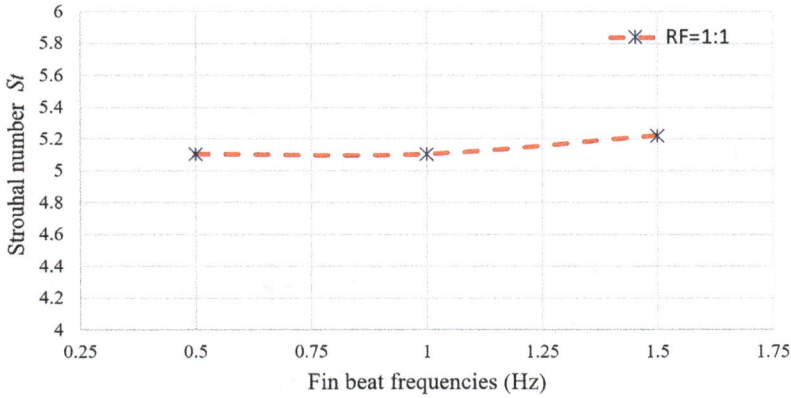

Figure 3.31. Strouhal number at $R_F = 3$.

Table 3.3. Comparison in terms of St (1 Hz).

References	St
This work	5.109
[10]$A = 60°$	7.76
[26] -JF1	6.05

equal to 2100, it indicates a laminar flow, and if it is greater than 4000, it indicates a turbulent flow. The higher the Reynolds number, the lesser the viscosity plays a role in the flow around the airfoil. Figure 3.32 shows the drag force generated by the concave fin for various values of Reynolds numbers. The generated drag force is proportional to the velocity in the case of laminar flow, and it is proportional to the square of the velocity at turbulent flow.

3.8.3. *Amplitude to Length Ratio*

Scaling parameters are of great importance in describing the dominant physics of locomotion. As mentioned earlier, efficient locomotion occurs when the value of the Strouhal (St) number is bound to the tight range of 0.2–0.4. On the other hand, another non-dimensional number that relates to fin-beat amplitude A, the length of the robot A/BL is bound in the range 0.1–0.3. However, the results in this study show a very close value to that range as shown in Figure 3.33.

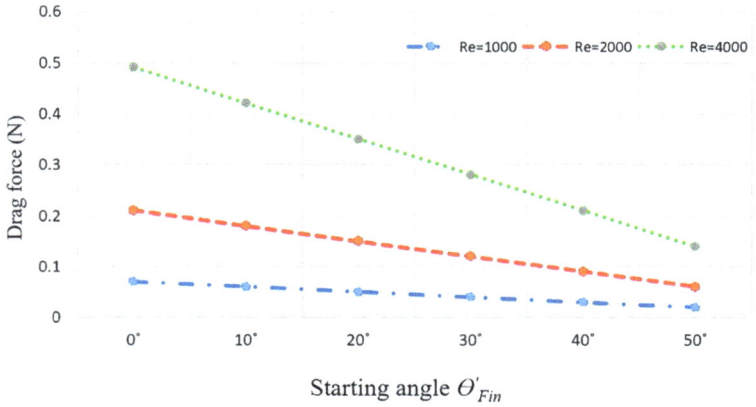

Figure 3.32. Starting angle of pectoral fin at different Reynolds numbers.

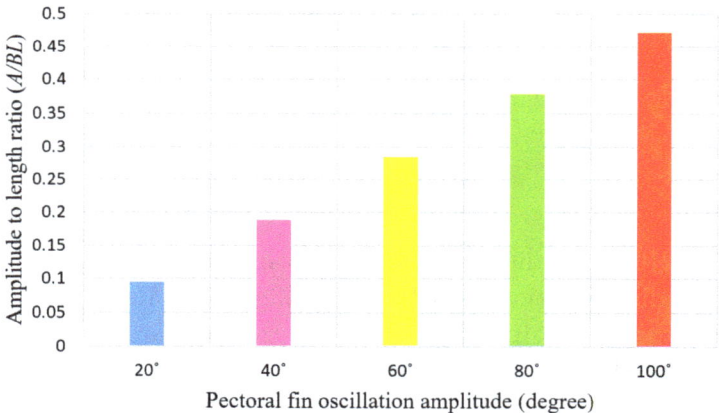

Figure 3.33. Amplitude to length ratio over different oscillation ranges.

3.9. Conclusion

In this chapter, a proposed model of a swimming robot actuated by a pair of concave pectoral fins has been presented. Several simulations and practical experiments are conducted, including studying the effect of the starting angle of the fin. The results showed that the best angle to start pushing in the power stroke phase is at $50°$. The effect of the oscillating amplitude range of fin is examined. Five ranges are studied, as presented in Chapter 2. The results showed a match between the results of the effects of the starting angle and increasing the oscillating range, so the greater the range, the greater the thrust. But due to the proposed shape of both the body and the fins being concave, the maximum range at angle 50° is sufficient to avoid collision with the frontal part of the head when the fin oscillates in higher ranges.

Then a transition was made to study the effect of frequency and it was noticed that there is a relationship between frequency and forward speed of the robot. At the low frequencies, this relationship is linear. That is, by increasing the frequency, the body speed increases for various values of R_F, but when increasing the frequency to the maximum, there is a clear effect of the shape of the fin with the values of R_F as it affects the body speed.

In terms of power to recovery ratio R_F, the results showed a great match between what was obtained in simulation and the results confirmed in practice. On the other hand, the efficiency of the proposed design was tested and compared to some real fish numbers like Strouhal number and amplitude to length ratio. On the other hand, the Reynolds number was dealt with as another factor in the proposed design. It has been tested at different starting angle values, exploring the extent of the effect of differing Reynolds numbers on the drag force generated at each angle.

Last but not least, the amplitude to body length ratio was studied and compared to the physical ranges of fish. In this case, the results obtained in this chapter showed a very large convergence between the real ratio of fish to the proportions of the proposed design of the robot and the various ranges of oscillation. This proves the efficiency of the suggested design as a prototype for an underwater swimming robot.

References

Arafat, H. N., Stilwell, D. J. and Neu, W. L. (2006). Development of a dynamic model of a small high-speed autonomous underwater vehicle, *in Proc. Oceans, Boston*, MA, pp. 1–6.

Aureli, M., Kopman, V. and Porfiri, M. (2010). Free-locomotion of underwater vehicles actuated by ionic polymer metal composites, *IEEE/ASME Transactions on Mechatronics*, 15(4).

Behbahani, S. B. and Tan, X. (2016). Bio-inspired flexible joints with passive feathering for robotic fish pectoral fins, *Bioinspiration & Biomimetics*, 11(3).

Behbahani, S. B. and Tan, X. (2017). Role of pectoral fin flexibility in robotic fish performance, *J. Nonlinear Sci.*, 27(4), pp. 1155–1181.

Eldredge, D. J. (2006). Numerical simulations of undulatory swimming at moderate Reynolds number, *Bioinspiration & Biomimetics*, 1(4), pp. 19–24.

Fedak, V., Durovsky, F. and Uveges, R. (2014). Analysis of robotic system motion in simmechanics and matlab GUI environment, *In Tech.*, pp. 1–19.

Fossen, T. I. (2011). *Handbook of Marine Craft Hydrodynamics and Motion Control*, 1st Ed., Norway, John Wiley & Sons Ltd., pp. 122.

Guerrero, J. E. (2010). Wake signature and Strouhal number dependence of finite-span flapping wings, *J. Bionic. Eng.*, 7, pp. 109–122.

Hatton, R. L., Burton, L. J., Hosoi, A. E. and Choset, H. (2011). Geometric maneuverability, *IEEE/RSJ International Conference on Intelligent Robots and Systems*.

Hernandez, R. D., Mora, P. A., Avilies, O. F. and Ferreira, J. (2016). Dynamic modeling and control PID of an underwater robot based on method hardware in the loop, *International Review of Mechanical Engineering*, 10(7), pp. 482–490.

Jagadeesh, P., Murali, K. and Idichandy, V. G. (2009). Experimental investigation of hydrodynamic force coefficients over AUV hull form, *Ocean Engineering*, 36, pp. 113–118.

Johan, L., Leeuwen, V., Cees, J., Voesenek, K. and Müller, K. (2015). How body torque and Strouhal number change with swimming speed and developmental stage in larval zebrafish, *J. R. Soc. Interface.*, 12.

Kanso, E. (2010). Swimming in an inviscid fluid, *Theoretical and Computational Fluid Dynamics*, 24, pp. 201–207.

Kunes, J. (2012). *Dimensionless Physical Quantities in Science and Engineering*, Elsevier Inc., pp. 1–443.

Li, N. and Su, Y. (2016). Fluid dynamics of biomimetic pectoral fin propulsion using immersed boundary method, *Applied Bionics and Biomechanics*, 2016, pp. 1–22.

Naser, F. A. and Rashid, M. T. (2019). Design, modeling, and experimental validation of a concave-shape pectoral fin of labriform-mode swimming robot, *Engineering Reports*, 1(5), pp. 1–17.

Naser, F. A. and Rashid, M. T. (2020). Effect of Reynold number and angle of attack on the hydrodynamic forces generated from a bionic concave pectoral fins, *IOP Conf. Ser.: Mater. Sci. Eng.*, 745, pp. 1–13.

Naser, F. A. and Rashid, M. T. (2020). The influence of concave pectoral fin morphology in the performance of labriform swimming robot, *Iraqi Journal for Electrical and Electronic Engineering*, 16(1), pp. 54–61.

Naser, F. A. and Rashid, M. T. (2021). Enhancement of labriform swimming robot performance based on morphological properties of pectoral fins, *J. Control Autom. Electr. Syst.*, 32, pp. 927–941.

Naser, F. A. and Rashid, M. T. (2021). Design and realization of labriform mode swimming robot based on concave pectoral fins, *Journal of Applied Nonlinear Dynamics*, 10(4), pp. 691–710.

Naser, F. A. and Rashid, M. T. (2021). Implementation of steering process for labriform swimming robot based on differential drive principle, *Journal of Applied Nonlinear Dynamics*, 10(4), pp. 737–753.

Naser, F. A. and Rashid, M. T. (2021). Labriform swimming robot with steering and diving capabilities, *Journal of Intelligent & Robotic Systems*, 103(14), pp. 1–19.

Rashid, M. T. and Rashid, A. T. (2016). Design and implementation of swimming robot based on labriform model, *Al-Sadeq International Conference on Multidisciplinary in IT and Communication Science and Applications (AIC-MITCSA)*, pp. 1–6.

Rashid, M. T., Naser, F. A. and Mjily, A. H. (2020). Autonomous micro-robot like sperm based on piezoelectric actuator, *International Conference on Electrical, Communication, and Computer Engineering (ICECCE)*, pp. 1–6.

Rohr, J. and Fish, F. E. (2004). Strouhal numbers and optimization of swimming by odontocete cetaceans, *The Journal of Experimental Biology*, 207, pp. 1633–1642.

Saadat, M., Fish, F. E., Domel, A. G., Di Santo, V., Lauder, G. V. and Haj-Hariri, H. (2017). On the rules for aquatic locomotion, *Physical Review Fluids*, 2.

Su, S., Jiang, Y. and Shi, O. (2014). Robotic fish's movement based on kalman filter and PID control, *Applied Mechanics and Materials*, pp. 568–570.

Ullah, B., Ovinis, M., Azhar, S., Khan, B. and Javaid, Y. (2020). Effect of waves and current on motion control of underwater gliders, *J. Mar. Sci. Technol.*, 25, pp. 549–562.

Chapter 4

STEERING PROCESS
OF SWIMMING ROBOT

4.1. Introduction

The control of orientation is related to the performance of the maneuver. Maneuverability is an important aspect of the kinematic performance of aquatic animals and underwater vehicles, such as swimming robots. Maneuvering is an integral part of steering, correcting path, obstacles avoidance in a complex environment, achieving stability in high-energy surroundings, jumping and diving, tracking targets, swimming through active turbulent waters. Maneuver studies focused on lateral steering (i.e. yawning). Aquatic animals show high performance of maneuverability with sacrificing not much in the stability. For example, a fish can achieve good speed while steering at a radius of 10–30% of its body length.

The narrow and fast rotation is distinguished by maneuverability and agility, respectively. Maneuverability is the ability to rotate in a narrow space and is measured as the radius specified by the length of the turning path (R/BL, where R is the turning radius and BL is the total length of the body). The maximum maneuverability is related to the minimum turning radius. In some cases for both AUVs (for example, six-finned BAUV), and animals (for example, Boxfish, rays, and squids), the bodies can rotate with a purely rotating, zero-turning radius. However, the area bounded by the rotating lengths of the rigid bodies of these swimmers still limits the occupied space during the maneuver (e.g. Boxfish, Ostracion Meleagris).

107

Flexible bodies are found in several animals and BAUVs, although executing non-zero radius steering is achieved in small spaces. A gliding robotic fish can rotate with a radius of 0.34 m (0.38 BL) using a tail fin tilted at 60°, but the glider with passive buoyancy was turned by a downward vortex. Therefore, the 3D helical maneuver was performed in a larger volume space than if the glider was kept at a constant depth.

It was previously observed that the contrast in maneuverability between fish and swimming robots was large. When comparing the maneuverability of swimming robots currently with animals as shown in Figure 4.1, it is evident that animals still generally showed advantages in maneuverability and showed the lowest turning radius with respect to body length compared to ergonomically designed devices. The turns are particularly wide for conventional swimming robots, which feature an inflexible torpedo-shaped body with few control surfaces typically placed in the rear of the body. The rotation of the small radius that is performed by animals results from the elastic body of many species and the many control surfaces they possess. In the case of a rigid boxfish, there are several moving fins in the front, back, and above the center of mass making the animal very maneuverable. The development of control systems and using flexible bodies in swimming robots have effectively decreased the steering radius compared to conventional automatic underwater vehicles (AUVs), but the current technology still lags the animals' ability to maneuver.

In contrast to maneuverability, agility is the speed at which orientation can be changed as an indicator of the rate of rotation and is measured as a change in angular velocity. When combined with maneuverability and speed, animals have a clear advantage in agility over current AUV and swimming robot technologies (see Figure 4.1). For animals, the rotation rate can be more than 5000 deg/sec, while conventional AUVs have a maximum rotation rate of 75 deg/sec, and swimming robot scores only reach 670 deg/seg. However, in general, agility decreases with body length and is limited by body flexibility as shown in Figure 4.1. Although, some marine mammals have large

Table 4.1. Turn rate for AUVs, swimming robots, and animals.

Type	Length (m)	Turn rate (deg/sec)	Reference
AUVs			
Generic UUV	6.1	4	Anderson & Chhabra (2002)
Microrobot	0.05	6	Ye *et al.* (2007)
REMUS	1.6	9.88	Stanway (2008)
USS Albacore	62.1	2	Miller (1991)
Swimming Robots			
Dolphin Robot	0.56	31.88	Yu *et al.* (2012)
Finnegan	2	53.6	Stanway (2008)
Fishlike Robot D	0.38	43.9	Pollard & Tallapragada (2019)
Four-Fin UUV	0.4	30	Geder *et al.* (2011)
MantaBot	0.4	53.7	Fish *et al.* (2017)
Multijoint Fish Robot	0.5	670	Su *et al.* (2013)
Multijoint Fish Robot	0.59	200	Su *et al.* (2013)
Robopike	0.82	17.5	Kumph (2000)
Robotic Fish	0.34	36	Hirata *et al.* (2000)
Robotic Fish	0.37	63.8	Yu *et al.* (2016a)
Robotic Fish	0.41	53	Yu *et al.* (2008)
Robotic Fish	0.41	62.51	Yu *et al.* (2008)
Robotic Fish	0.5	200	Su *et al.* (2013)
Robotic Fish/Pectoral Fins	0.15	9	Behbahani *et al.* (2013)
Robotic Killer Whale	1.58	7.8	Wu *et al.* (2019)
Robotuna	2.4	75	Anderson & Chhabra (2002)
Six-finned BAUV	0.91	40	Menozzi *et al.* (2008)
Wire-Driven Robot Fish	0.31	6.84	Li *et al.* (2013)
Animals			
Amazon River Dolphin	2.56	123.75	Fish (2002)
Angelfish	0.07	3244.08	Domenici & Blake (1991)
Beluga Whale	3.23	245	Fish (2002)
Bottlenose Dolphin	2.36	405	Fish (2002)
Boxfish	0.12	200.5	Walker (2000)
Brief Squid	0.06	302.6	Jastrebsky *et al.* (2017)
Brief Squid	0.09	725.8	Jastrebsky *et al.* (2016)
California Sea Lion	1.72	690	Fish *et al.* (2003a)
Commerson's Dolphin	1.28	366	Fish (2002)
Crocodile	1.1	230.4	Frey & Salisbury (2001)

(*Continued*)

Table 4.1. (*Continued*)

Type	Length (m)	Turn rate (deg/sec)	Reference
Cuttlefish	0.09	383	Helmer *et al.* (2017)
Cuttlefish	0.03	485	Jastrebsky *et al.* (2016)
Dasyatis Ray	1.05	32	Parson *et al.* (2011)
False Killer Whale	3.55	252.5	Fish (2002)
Hammerhead Shark	0.75	1221	Kajiura *et al.* (2003)
Harbor Seal	1.7	388.7	Geurten *et al.* (2017)
Killer Whale	5.05	232.5	Fish (2002)
Leopard Shark	0.35	300.23	Porter *et al.* (2011)
Manta Ray	1.25	67.32	Fish *et al.* (2018)
Myliobatis Ray	0.82	48	Parson *et al.* (2011)
Painted Turtle	0.05	501.8	Rivera *et al.* (2006)
Penguin	0.58	575.8	Hui (1985)
Rainbow Trout	0.39	2332.75	Webb (1976)
Rainbow Trout	0.26	3114.143	Webb (1983)
Rainbow Trout	0.1	5157	Webb (1976)
Sandbar Shark	0.69	246.37	Kajiura *et al.* (2003)
Smallmouth Bass	0.24	5509.61	Webb (1983)
Spiny Dogfish	0.59	1221	Domenici *et al.* (2004)
Steller Sea Lion	2.26	274.27	Cheneval *et al.* (2007)
Striped Dolphin	2.13	453.33	Fish (2002)
Turtle (*Chrysemys*)	0.13	397.5	Mayerl *et al.* (2018)
Turtle (*Emydura*)	0.18	477.8	Mayerl *et al.* (2018)
Whirigig Beetle	0.01	4437.5	Fish & Nicastro (2003)
Yellowfin Tuna	0.35	425.6	Blake *et al.* (1995)

sizes, (for example, whales, dolphins, seals, and sea lions) they exhibit high turnover rates compared to AUVs and swimming robots of equivalent or smaller size.

This chapter aims to introduce two studies on the steering strategies that are used in Labrifom mechanism mode. These two strategies include the study of the rigid body of boxfishes being able to achieve high performance in the steering process based on the caudal fin, while the other study includes the modeling and experimental validation of this model to achieve the steering process by pectoral fins.

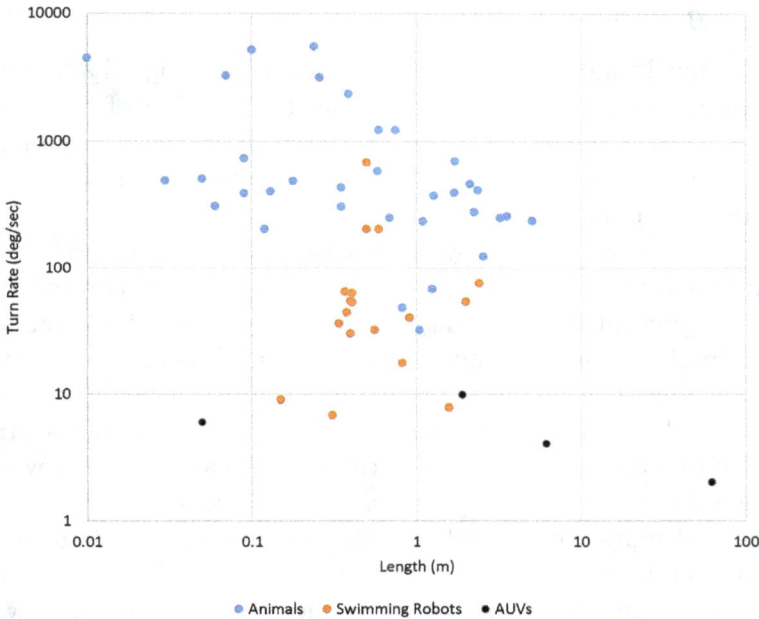

Figure 4.1. Comparison of turn rates for AUVs, swimming robots, and animals with respect to body length.

4.2. Steering by a Caudal Fin in the Boxfish

Although the boxfish has rigid armor that limits body undulation, it is highly maneuverable and can swim with remarkable dynamic stability. In this section, the steering process using the caudal fin in boxfish will be studied. Turning torque in the flow basin is measured using a physical model with attachable caudal fin closed or open at the different body and tail angles, and different water flow velocities, as the results are evaluated by the simulated computational fluid dynamics (CFD), which indicates that the caudal fin achieves steering. The caudal fin acts as a rudder for a rigid body that is naturally unstable concerning steering. Boxfishes appear to use unstable body interaction and active changes in the shape and direction of their caudal fin to modify their maneuverability and stability.

4.2.1. *Background*

It is assumed that a trade-off exists in the morphological adaptations of locomotion depending on the ecological position of the species. This will result in the formation of a body specialized in maneuverability with a good performance in this function but not a superior performance in all cases.

The stiffness imposed by the boxfish armor has important consequences for thrust, stability, and maneuverability. For swimming and slow upright maneuvering, boxfishes use different combinations of pectoral, dorsal, and anal fins for thrust, using the dominant pectoral/anal fins and the dominant anal/dorsal fins. The ability of a boxfish to steer is of great importance for a boxfish's fitness. Despite this unique bauplan, box thicknesses can perform low recoil movements with a turning radius close to zero. This ability to rotate and maneuver is perfectly in line with the requirements for locomotion in the complex three-dimensional (3D) coral habitat where boxfishes live. However, the question remains on how can boxfish be highly maneuverable without compromising the stability needed for controlled and effective swimming.

The close relationship between the angle at which the caudal fin bends to the side and the radius of rotation indicates that the boxfish, like most other fish, controls rotation by engaging the caudal fin as a rudder. Several studies have assumed that in a wide range of fish species the caudal fin can be used for steering and that boxfishes may partially control their movement using the caudal fin as a rudder. The caudal fin might play an important role in mitigating this instability, and it is calculated that a caudal fin with a surface area of $0.001\,\mathrm{m}^2$ would be able to withstand destabilizing torques of the object itself. Thus, the caudal fin is assumed to be important in spin maneuvers and thus in controlling instability. In addition, the condition of the caudal fin (completely closed or open) and the angle with respect to the body axis is expected to be important during rotation maneuvers according to the correlations between the tail angle and radius of rotation. During slow forward swimming, the caudal fin remains closed and is aligned along the middle axis of the boxfish, while the caudal fin is opened during fast forward swimming. The caudal fin is closed when rotating, and the tail rotating in the opposite direction compared to the body.

This study aims to determine the effect of the caudal peduncle and the caudal fin on the unstable deflection properties of the body of boxfish. Two physical models of boxfish were constructed to resolve these effects. The first model is a reconstructed rigid caudal brace while in the other model, the caudal abutment could be attached and rotated at different angles with respect to the body. In a flow pool, steering torques are measured at different velocities of water flow, further at several steering angles. Firstly, both models are tested in case of neglecting the caudal fin to quantify the effect of adding only a caudal peduncle to the carapace. After that, the models are provided with either closed or open reconstructed caudal fin. In addition, the caudal peduncle and the caudal fin are installed at several angles with respect to the body axis to observe the hydrodynamic interaction. This hydrodynamic interaction is particularly interesting due to the wide shield of the boxfish, which may protect the caudal fin from the incoming flow, and the aforementioned eddies thrown by the shield under certain angles of attack, which may create the torque effect by these eddies. The CFD has been used to validate the physical model data and model the stresses and torque exerted on the body, the caudal peduncle, and the caudal fin.

4.2.2. *Model, Caudal Peduncle and Caudal Fin Reconstruction*

Laser scanning surface is used to produce two physical models produced based on a fused-deposition modeling 3D printing technique (RapMan 3.1 3D Printer Kit, Bits from Bytes, 3D Systems Inc., Rock Hill, USA). Surface patches are removed with a plaster-based filler (Alabastine, AkzoNobel) and subsequent sanding. A hole is drilled in the ventral side of the models in which a threaded attachment is fitted in such a way that the cardiac line for insertion is in line with the center of the model size. The models are spray-painted with a black waterproof matte finish (Motip EAN 8711347040018). Assume that the center of mass is at the center of the volume, which could lead to some errors in the evaluation of the relevant steering torque.

A previous study of experimentally measured mass centers and center of volume based on surface shields showed that the deflections are limited to a small percentage of the shield length.

In order to reconstruct the caudal peduncle and caudal fin of these two models, a living boxfish is filmed with a length of 15.5 cm swim in the laboratory with seawater pool at the University of Groningen by a camera at 50 fps (Adimec 1000 m, 1.2 Mpix, Adimec Eindhoven, The Netherlands; Matrox Solios v. 1.0 frame grabber board, Matrox Imaging, Durval, Quebec, Canada) supported by real-time software for snapping images (Ingenico, Lage Zwaluwe, The Netherlands). The caudal peduncle is reconstructed based on the physical models by using Aquascape Aquarium Epoxy (D-D The Aquarium Solution Ltd) and multiple photographs as reference material. In the first model, the caudal peduncle is reconstructed which is modified focusing on the midline of the model as in Figures 4.2(a) and 4.2(b). In the second model, the caudal peduncle is attached to the body shield by a joint, allowing it to turn and be fixed at several angles as in Figures 4.2(c) and 4.2(d).

A side view of live boxfish in the case of caudal fin fully open and closed was filmed in order to construct the caudal fin. The caudal fin profiles were determined using image 1.42q and measured isometrically at a few percent to achieve a size identical to the shield lines from a side view. The closed caudal fin and open caudal fin have been selected on the basis of the lowest and highest surface area

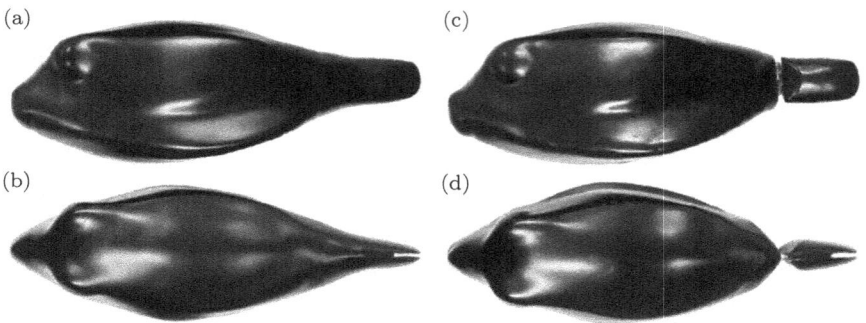

Figure 4.2. Photographs of the physical models of boxfish. (a) Lateral and (b) dorsal view of the rigid tail model. (c) Lateral and (d) dorsal view of the turnable tail model.

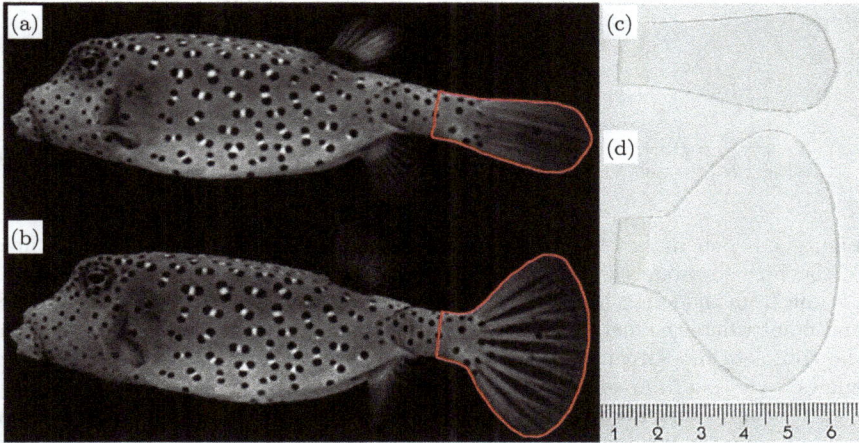

Figure 4.3. Photographs of the living boxfish specimen in lateral view with (a) closed caudal fin and (b) open caudal fin from which the contours were extracted (red line). These contours are isometrically scaled to construct the (c) closed caudal fin ($8.226\,\mathrm{cm}^2$) and (d) open caudal fin ($19.828\,\mathrm{cm}^2$) which were attached to the physical models of boxfish. Scale bar is in centimeters and only applicable to the right panels.

among at least five images, respectively as shown in Figures 4.3(a) and 4.3(b). The caudal fin models are cut from sheets of poly (methyl 2 methyl propanoate) (PMMA) 1 mm thick with Young's modulus of approximately 3 GPa and sanded at the edges as shown in Figures 4.3(c) and 4.3(d). Because this material is relatively rigid but still bends slightly at greater forces, it has been considered a reasonable imitation of a caudal fin, although species are commonly observed to form fins due to the interaction with water and the geometry of the fin rays and properties of the fin material, such as cupping, cannot be included in our simplified model. The vertical saw cut in the middle of the tail supports the physical models of both fish, allowing the tail fins to be attached to the models interchangeably using kneading glue (Poster Buddies, Pritt, Henkel AG & Co., Germany). The tail support of the reversible tail model can be fixed at any angle of attack (as angle θ that is shown in Figure 4.4) with respect to the midline of the model using kneading glue to fill the gap around the hinge and strengthen the copper wire to prevent deformation during an experiment. The anterior surface

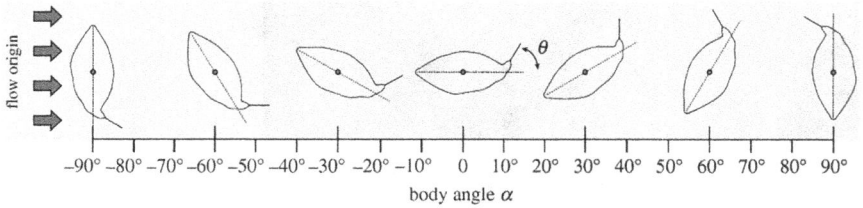

Figure 4.4. Schematic of the dorsal view of the experimental set-up. The outline of the boxfish models is shown. The arrows indicate the direction of the flow, coming from the left. The small circle shows the position of the center of volume and point where the metal rod is attached to the model, hence the point at which the complete model could be rotated (fulcrum). The models are fixed at body angles α varying from $-90°$ up to $90°$ with $10°$ increments. The $-90°$ to $0°$ range and $0°$ to $90°$ range are measured in separate series. The rigid tail model had the tail permanently fixed at an angle θ of $0°$ relative to the midline axis of the body (dotted line). The turnable tail model had the tail fixed at the angles θ of $0°$, $10°$, $20°$, $30°$ and $40°$. Note that positive tail angles θ correspond to deflection in the same, counterclockwise direction as positive body angles α.

areas are 21.7 and 21.1 cm^2 for the hardtail and rotatable tail models. Both models are 0.16 dm^3 in size and 16.95 cm long with a caudal fin. The side view surface area of the closed caudal fin is 8.23 cm^2 and the open caudal fin is 19.83 cm^2, respectively.

The 3D laser scanning CFD model is modified with the addition of a caudal stalk and a caudal fin according to the above procedures. The caudal peduncle and caudal fin of the physical models are mimicked using numerical transformations of the outer contour features based on the fishes imaged. The triangular surface is transformed into a smooth, rational irregular B-Spline surface (NURBS, VRMesh 5.0, VirtualGrid, Bellevue, WA, USA) (see Figure 4.5), enabling the computation of hydrodynamic pressures and torque exerted separately on the body, caudal peduncle, and caudal fin. In all CFD simulations, the space between the body and the caudal peduncle is not filled as is done in the physically rotatable tail model.

4.2.3. *Specification of Environment Measurements*

Steering torque measurements are performed in a 300 L flow tank containing pure water at 20°C. The dimensions of a flow tank with a laminar flow are 25 × 25 cm in width and height, while 50 cm in

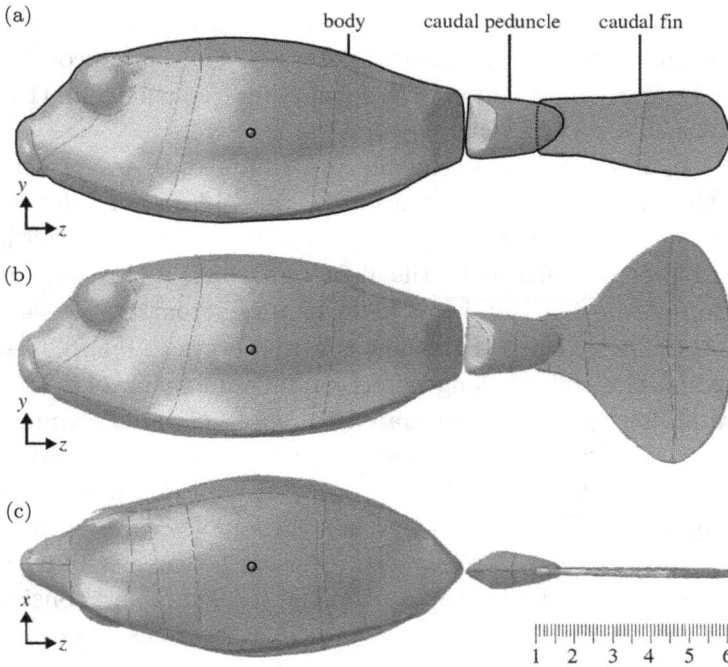

Figure 4.5. Geometry of the boxfish models used in computational fluid dynamics (CFD) simulations. (a) Lateral view of the model with closed caudal fin. (b) Lateral view of the model with open caudal fin. (c) Dorsal view of the model. The small circle shows the position of the center of volume. The subdivision of the boxfish's surface in body, caudal peduncle and caudal fin is indicated at the top. Scale bar is in centimeters.

length. In order to prevent wave formation, the transparent PMMA plate covers the top of the test section. For measurements, the models on a 5 mm stainless steel rod are mounted attached to a measuring platform placed on the tank, where the center of the model size is exactly in the middle of the tank cross-section.

The rod is installed in a specially made ball bearing seat, acting as a pivot that could rotate freely and is practical without friction. After adjusting the angle of the model with respect to the flow (body angle α; see Figure 4.4), torque can be measured as a function of flow velocity using a 40 mm arm mounted to the pivot by a force sensor (Vernier Dual range 10 N/50 N, DFS-BTA, Vernier Software

& Technology, Beaverton, USA) which is connected to a personal computer using the Vernier LabPro interface. Data are recorded with Vernier LoggerPro 3.8 software at a sampling frequency of 50 Hz after calibration.

In the first set of experiments, no caudal fins are attached to the boxfish models and the tail of the turnable tail model is fixed at an angle θ of $0°$ with respect to the midline axis of the model body. Models are placed in the flow tank with different angles of deflection of the α body, i.e. $0°–90°$ in $10°$ increments as shown in Figure 4.4. The velocity of the flow is varied between 0 and 0.5 m/sec (approximately 3.5 body lengths/sec) in increments of 0.1 m/sec. The maximum flow velocity is estimated to represent rapid swimming in relatively large individuals. These velocities also roughly correspond to the measured flow velocities in coral reef habitats, which interfere with the distribution of the yellow box thickness according to Myers. Five iterations of each velocity are recorded for each α body angle. One recording lasted 10 sec of continuous recording and consisted of 501 data points. Finally, the mean of the five iterations was calculated and the standard deviations were calculated.

Next, experiments are repeated with both boxfish models, which includes an artificial closed caudal fin and an open caudal fin fixed at an angle θ of $0°$ (see Figure 4.4). Measurements are recorded at body angles α of $0°$ to $60°$ with $10°$ increments. Reliable measurements at angles higher than $60°$ are not possible due to wall effects because the caudal fin is too close to the wall of the flow tank; The minimum distance is maintained at 15 mm which is confirmed by particle image velocity measurement with no effects on the wall. Finally, experiments are repeated with the turnable tail model with the tail placed at different angles θ (see Figure 4.4), respectively $10°$, $20°$, $30°$, and $40°$ with closed and open caudal fins, applying the same measurement procedure as described above. The ranges of $90°–0°$ and $0°–90°$ are measured in a separate series for tail angles of $10°$, $20°$, $30°$, and $40°$, resulting in doubling the number of data points for α body angles of $0°$. For the rigid tail model experiments and the tail model turnable with a tail angle θ of $0°$, the steering torque increasing the angle between the flow and the model's midline is determined as positive. Steering torque is defined as positive when operating in a counterclockwise direction from a dorsal point of view on the boxfish (see Figure 4.4). For $\alpha > 0$, the positive torque would of course destabilize, i.e. increase the angle between the flow and the midline in the model. For $\alpha < 0$, the positive torque stabilizes the path, i.e. reduces the angle between the flow and the midline of the model.

4.2.4. *Computational Fluid Dynamics (CFD) Simulations*

The CFD surface model is imported into ANSYS Design Modeler 14.5.7 (ANSYS Inc., Canonsburg, PA, USA) and placed within a cylindrical outer boundary of the flow field. The angles of the body and tail are determined as shown Figure 4.5. The flow domain is integrated into the ANSYS 14.5.7 network and imported into ANSYS Fluent 14.5.7, where the boundary conditions are set: the size of the grid elements, the position of the boxfish model in the flow domain, the external boundary dimensions of the flow domain. All torques resulting from CFD analysis are based on both shear and compression forces. CFD allows the subdivision of forces and torque to the body, the caudal peduncle, and the caudal fin to assess its relative contribution to total torque. In the first set of CFD measurements, the model was placed at body angles α of 60°, 40°, 20°, 0°, 20°, 40° and 60° with a tail angle θ of 20° with the caudal fin closed and open caudal fin at a flow velocity of 0.5 m/sec to verify experimental data of the flow tank (see Figure 4.4). Second, the CFD model was fixed at an α body angle of 20° with tail angles varying θ by, 60°, 40°, 20°, 0°, 20°, 40°, and 60° as in Figure 4.4.

4.2.5. *Results*

4.2.5.1. *Rigid tail model and turnable tail model: No caudal fin at tail angle θ of 0°*

The steering torque about the center of the volume relative to the angle of the body show a similar pattern for both the solid tail model and the turnable tail model when no caudal fin is attached as shown in Figure 4.6. In the case of angle increases from its original direction in line with the flow (body angle α of 0°), the steering torque around

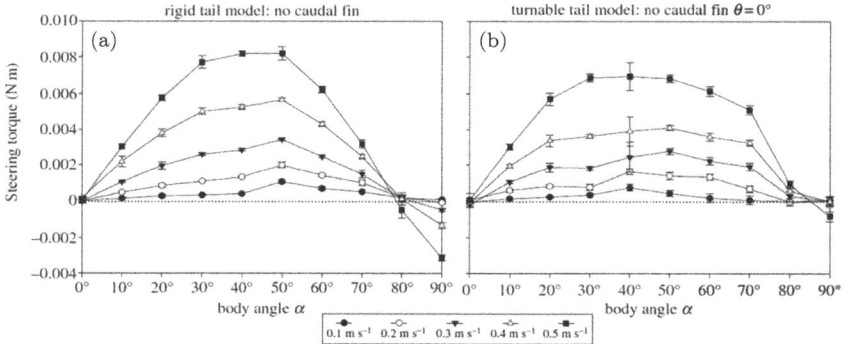

Figure 4.6. Performance of the (a) rigid tail model and (b) turnable tail model (tail angle θ of $0°$) without caudal fin. Steering torque (N.m) about the center of volume per body angle α ($°$) for boxfish. Experimental data are given for five flow speeds between $0.1\,\mathrm{m/sec}$ and $0.5\,\mathrm{m/sec}$ with $0.1\,\mathrm{m/sec}$ increments. The dotted lines indicate a net steering torque of $0\,\mathrm{N.m}$. Error bars shown for all flow speeds denote the between-repeat standard deviation.

the center of the body volume becomes increasingly positive (i.e. the direction of rotation away from the flow or instability) until the peak of the rigid tail model is reached at $40°$ and $50°$ (flow tank at $0.5\,\mathrm{m/sec}$) or $50°$ (flow tank at $0.4\,\mathrm{m/sec}$) (see Figure 5.6(a)), and for tail model turnable at either $30°$ and $40°$ (flow tank at $0.5\,\mathrm{m/sec}$) or $50°$ (the flow tank up to $0.4\,\mathrm{m/sec}$) (see Figure 4.6(b)). For larger body angles, the steering torque is decreased but remains disturbed up to an angle of $80°$ (see Figure 4.6). For the rigid tail model and to a lesser degree for the turnable tail model, the torque becomes negative (i.e. direction of rotation towards flow or stabilizing) with a body angle of $90°$. Larger flow velocities generate more torque.

4.2.5.2. *Rigid tail model and turnable tail model: Closed caudal fin and open caudal fin at tail angle θ of $0°$*

The steering torque about the center of volume relative to the angle of the body shown a pattern similar to that of the solid tail model and the turnable tail model with a closed caudal fin and an open caudal fin with a tail angle of $0°$ as shown in Figure 4.7. When the body angle increases from its original direction in line with the flow (body angle α of $0°$) for both models with a closed tail fin,

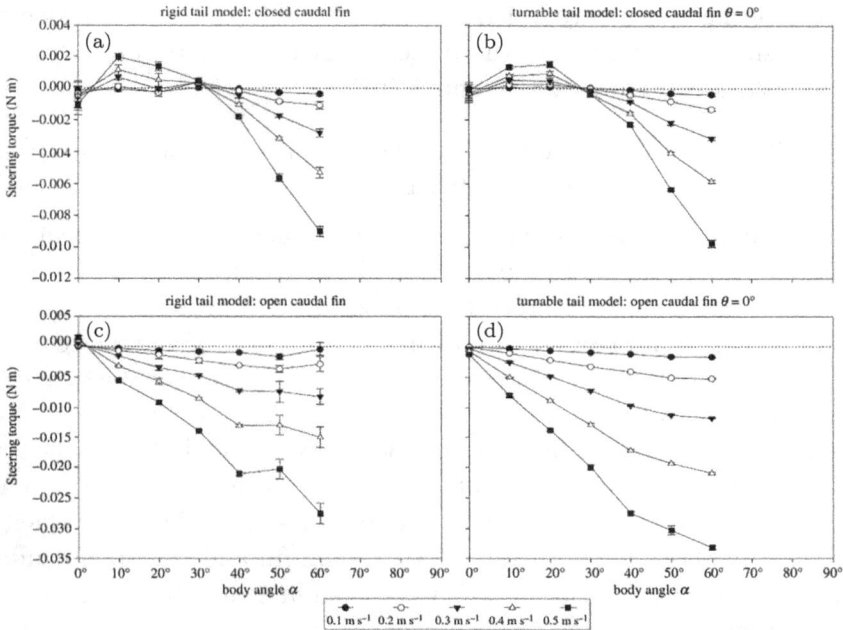

Figure 4.7. Performance of the (a, c) rigid tail model and (b, d) turnable tail model with a (a, b) closed caudal fin and (c, d) open caudal fin at a tail angle θ of $0°$. Steering torque (N.m) about the center of volume per body angle α (°) for boxfish. Experimental data are given for five flow speeds between 0.1 and 0.5 m/sec with 0.1 m/sec increments. The dotted lines indicate a net yaw torque of 0 N.m. Error bars shown for all flow speeds denote the between-repeat standard deviation.

the steering torque around the center of the body volume becomes increasingly positive (i.e. the direction of rotation away from the flow or instability) to reach the peak of the tail model annealing at $10°$ (flow tank at all velocities) (see Figure 4.7(a)) and at $20°$ for turnable tail model (flow tank at all velocities) (see Figure 4.7(b)). At larger body angles, torque decreases to 0 N.m at a body angle of about $30°$ for both models and becomes increasingly negative with increasing body angles (i.e. direction of rotation towards flow or fixation) for both models (all velocities) (see Figures 4.7(a) and 4.7(b)). When the body angle increases from its original direction in line with the flow (body angle α of $0°$) for both models with an open caudal fin, the steering torque around the center of the body volume becomes increasingly negative (i.e. the direction of rotation towards flow or stability) (see Figures 4.7(c) and 4.7(d)).

4.2.5.3. *Turnable tail model: Closed caudal fin and open caudal fin at tail angles θ of 0°, 10°, 20°, 30° and 40°*

The effect of closed caudal fin orientation and open caudal fin on the model boxfish at the different body and tail angles is shown in Figure 4.8. The y-axes are the same scale in all graphs to facilitate comparison. The steering torque about the center of volume relative to the angle of the body show a pattern similar to that of the turnable tail model with tail angles θ of 0°, 10°, 20°, 30°, and 40° for a closed caudal fin (Figures 4.8(a), 4.8(c), 4.8(e), 4.8(g) and 4.8(i)) And open caudal fin (Figures 4.8(b), 4.8(d), 4.8(f), 4.8(h) and 4.8(j)). The torque with respect to the tail angle becomes greater as the velocity of flow increases for both the closed caudal fin and the open caudal fin at all angles of the body α. Open caudal fin generates more torque than closed caudal fin. The data for the steering torque of the tail model rotating with a tail angle θ of 0° are the same as those in Figures 4.7(b) and 4.7(d). However, for comparison purposes, the data are flipped on the y-axis and the x-axis of zero (light gray area).

When the caudal fin is closed, the switching point between the left and right rotations (i.e. the point at which the torque is equal to 0 N.m, i.e. neutral; determined by the linear interpolation between the sample points with positive and negative torques) toward a body has more negative angles when increasing the tail angle for all velocities of flow (i.e. from a specific body angle α by 27° with a tail angle θ of 0°, $\alpha = 57°$ at $\theta = 10°$, $\alpha = 57°$ at $\theta = 20°$, $\alpha = 62°$ at $\theta = 30°$, to $\alpha = 68°$ at $\theta = 40°$; Figures 4.8(a), 4.8(c), 4.8(e), 4.8(g) and 4.8(i)). The torque relative to the angle of the object becomes greater with the increase in the flow velocity (maximum values of 0.014 and 0.018 N.m at a flow velocity of 0.5 m/sec), although the steering torque does not necessarily increase on increasing the angles of the object. When the caudal fin is opened, the switching point between the left and right rotations (i.e. the point at which the torque is equal to 0 N.m, i.e. neutral) shifts towards more negative

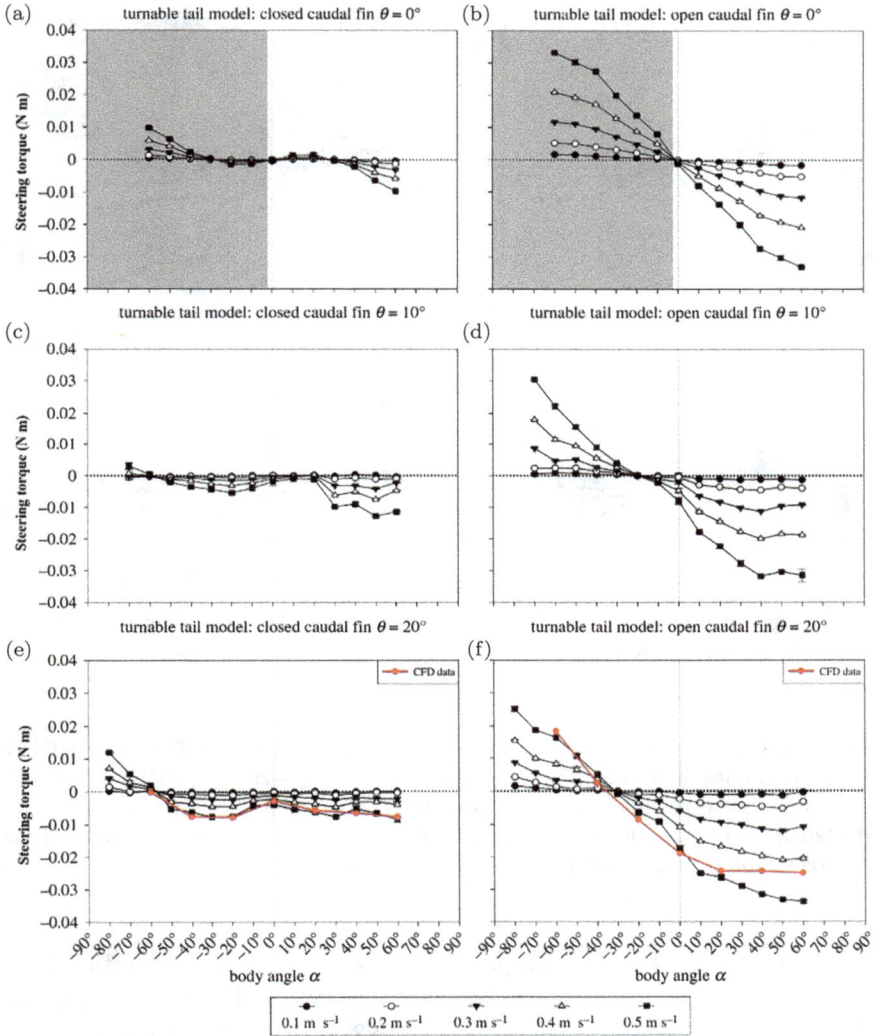

Figure 4.8. Performance of the turnable tail model with a (a, c, e, g and i) closed caudal fin and (b, d, f, h and j) open caudal fin at the tail angles θ of 0°, 10°, 20°, 30° and 40°. Steering torque (N · m) about the center of volume per body angle α (°) for boxfish. Experimental data are given for five flow speeds between 0.1 and 0.5 m/sec with 0.1 m/sec increments. (a) and (b) are based on data Figures 4.7(b) and 4.7(d)). The left-hand grey areas in (a) and (b) indicate mirrored data over the y-axis and x-axis from the right-hand white area. The y-axes are of the same scale in all graphs for easier comparison. The vertical grey lines indicate a body

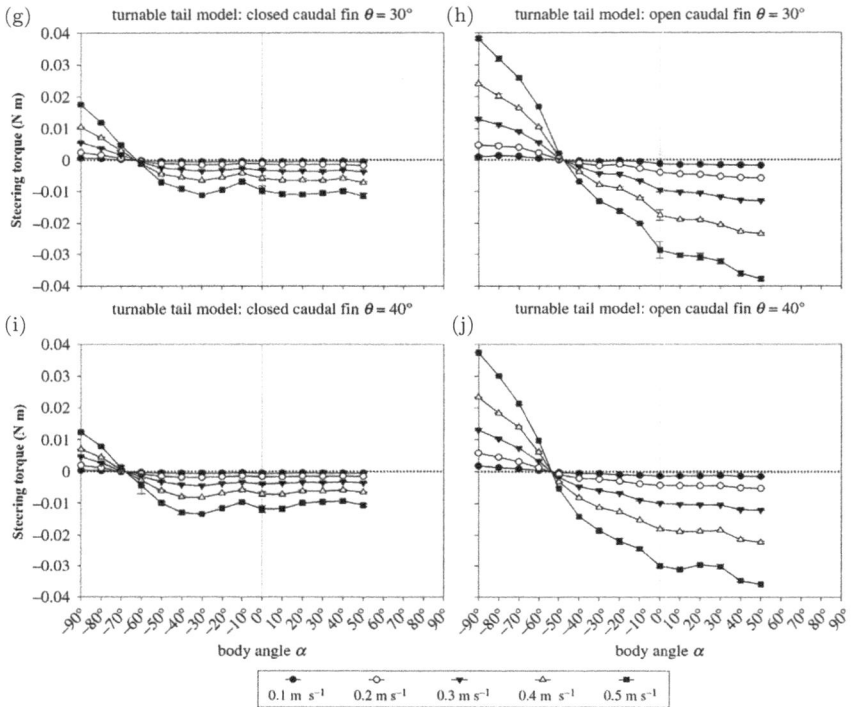

Figure 4.8. (*Continued*) angle α of $0°$. The red lines in (e) and (f) show the CFD simulation data for the body angles α of $-60°$, $-40°$, $-20°$, $0°$, $20°$, $40°$ and $60°$ at a tail angle θ of $20°$ with a (e) closed or (f) open caudal fin at $0.5\,\text{m/sec}$. The dotted lines indicate a net yaw torque of $0\,\text{N·m}$. Error bars shown for all flow speeds denote the between-repeat standard deviation.

body angles as the tail angle increases for all flow velocities (i.e. from a set the body angle α by $0°$ at tail angle $\theta = 0°$, $\alpha = 20°$ at $\theta = 10°$, $\alpha = 32°$ at $\theta = 20°$, $\alpha = 48°$ at $\theta = 30°$, to $\alpha = -54°$ at $= 40°$; Figures 4.8(b), 4.8(d), 4.8(h) and 4.8(j)). The torque with respect to the angle of the object becomes greater with the increase in the flow velocity (maximum values of 0.038 and 0.038 N.m at a flow velocity of $0.5\,\text{m/sec}$), although the steering torque does not necessarily increase when increasing the angles of the body. The torque data of the CFD simulation are in agreement well with the experimental results (Figures 4.8(e) and 4.8(f): red lines).

4.2.5.4. *Hydrodynamic forces on the body, caudal peduncle and caudal fin*

The pressure distribution on the yellow boxfish model with a body angle α and a tail angle θ of 20° illustrates the difference between the closed caudal fin and the open caudal fin as shown in Figure 4.9. When the caudal fin is closed, the area of high positive pressure on the flow-facing side of the caudal fin is much smaller than when the caudal fin is open, regardless of the tail angle θ (see Figures 4.9 and 4.10). Therefore, the torque produced by the caudal fin is greater when open compared to closed. Depending on the total torque of the model, resulting from the distribution of the applied pressure as shown in Figure 4.10, the model will shift to the left or the right. In the example shown in Figure 4.9(a), the net torque is 0.006 N.m (see Figure 4.8(e)), i.e. roughly neutral but only right-turning, while in Figure 4.9(b), the net torque is 0.026 N.m (see Figure 4.8(f)), i.e. right-turning towards the oncoming flow. The pressure distribution

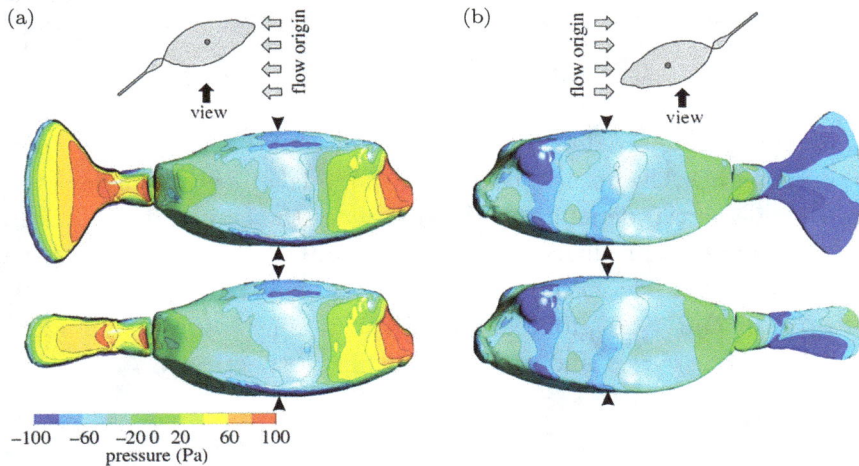

Figure 4.9. CFD simulations showing exerted pressure on (a) the flow-facing side and (b) the non-flow facing side of boxfish with open caudal fin (middle panels) and closed caudal fin (bottom panels) at a flow speed of 0.5 m/sec. The body angle α and tail angle θ are both 20°. The small circle in the grey line drawings at the top shows the position of the center of volume in dorsal view. The black arrowheads show the position of the axis of rotation, i.e. going through the center of volume.

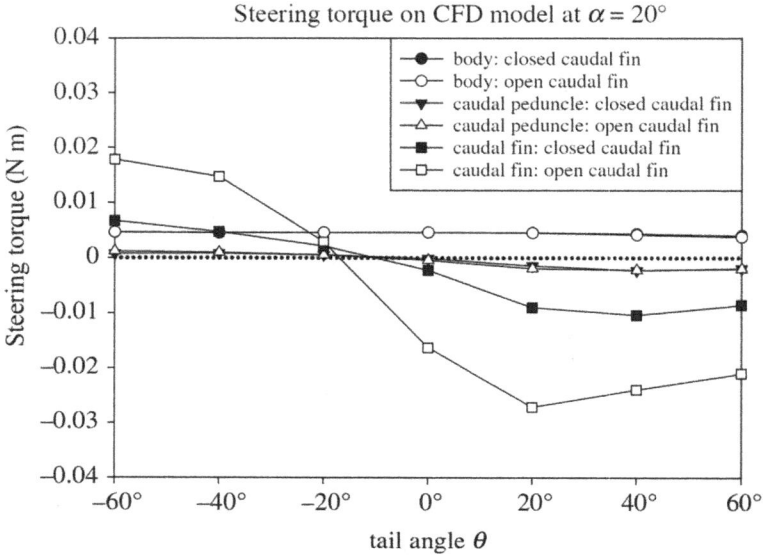

Figure 4.10. CFD simulation data showing the individual yaw torques on body, caudal peduncle and caudal fin of the boxfish model with a closed or open caudal fin at a flow speed of 0.5 m/sec. The body angle α is 20° and the tail angles θ are $-60°$, $-40°$, $-20°$, $0°$, $20°$, $40°$ and $60°$. Both lines of the body and caudal peduncle overlap. The dotted line indicates a net yaw torque of 0 N.m.

on the body does not depend on the condition of the caudal fin (closed or open). Thus, the net torque in the fish model at a given angle of the body and the tail depends on the torque by the caudal fin, which in turn depends on its surface area (Figures 4.9 and 4.10). The relative effect of the caudal prop on torque is negligible (Figures 4.9 and 4.10) at $\alpha = 20°$ and $= 20°$, for example, the clamping torque is only 20% and 7% of the closed and open caudal fins, respectively.

4.2.6. *Discussion and Conclusion*

Boxfish live in tropical coral waters and maneuver in crevices and around coral reefs, rocks, and sandy patches that feed on a variety of benthic organisms. Hence, it has been hypothesized that its morphological features enhance maneuverability. Several researchers have shown that the caudal fin could play an important role in controlling steering in boxfishes.

The results of the present study show that the caudal fin is indeed a major means of hydrodynamically unstable body fixation of boxfish (Figures 4.6 and 4.7). It has been hypothesized several times that the caudal fin plays a central role in controlling rotational maneuvers in many species of fish as well as in boxfishes. This cycle of stability has been demonstrated by measuring torque at different body angles (α) in a physical model flow tank of boxfish with a caudal peduncle and caudal fin. The results confirm the measurements and calculations made by Van Wassenbergh *et al.* (see Figure 4.6(c)). Both the rigid tail model and the turnable tail model are unstable without a caudal fin, which conforms to the model used by Van Wassenbergh *et al.* Greater torque appears destabilizing at higher flow velocities (see Figure 4.6). Moreover, the torque generated for both models is very similar as shown in Figure 4.6. Since the caudal peduncle was present in both the rigid tail model and the reversible tail model when a caudal fin is not attached, it can be concluded that the caudal peduncle alone does not have a significant effect on the stability of the boxfish trajectory, which was confirmed by CFD measurements (see Figure 4.10).

Results of experiments demonstrate an interaction between a destabilizing body and an anchored pathway. When a closed caudal fin is attached to either of the two models with a caudal angle θ of $0°$, an unstable torque is likely to occur at body angles of $10°$ and $20°$ as a result of this interaction as shown in Figures 4.7(a) and 4.7(b). The body has a greater effect at lower body angles than the closed caudal fin, and the latter becomes more dominant when the angle of the body increases. When the caudal fin is open, the steering torque is locked in its path at all body steering angles (see Figures 4.7(c) and 4.7(d)) indicating that the open caudal fin is dominant over the destabilizing net body. Thus, the opening or closing of the caudal fin significantly changes the nature of the torque balance due to the flow of water over the boxfish.

The same direction can be distinguished from experiments in which the caudal peduncle and the caudal fin have been fixed at different angles of the tail (the turnable tail model) and at different angles of the body with both the closed caudal fin and the open caudal fin (see Figure 4.8). These results demonstrate that boxfish is able to rotate using the interaction between the destabilizing body and the caudal fin stabilizing the path by opening or closing the caudal fin and rotating the tail. When the caudal fin is closed, the body has a more dominant effect on the total steering torque. When the caudal fin is open, the destabilizing effect of the body is strongly resisted. For illustration, when the tail angle is 20°, the model switching point with an open caudal fin is about the body angle α of 32° as shown in Figure 4.8(f). This is the net effect of the total body and the caudal fin. This switching point would theoretically be expected to have α body angle of 20° when the tail angle is θ of 20° if the body were to be hydrodynamically neutral. However, the body does cause destabilizing torque (e.g. 0.0056 N.m with α body angle of 20° and flow velocity of 0.5 m/sec as in Figure 4.6(b); note that negative body angle values α are not displayed but are expected to follow the same pattern, like positive body angles but reflected over the x-axis). This torque is approximately the same at an α body angle of 20° when the tail angle is θ of 20° with an open caudal fin (see Figure 4.8(f)). When the caudal fin is closed, the net unstable body is more dominant at smaller body angles because the caudal fin is located behind the body. At larger body angles, the tail adds more diffraction since the caudal fin extends from behind the body to the unperturbed water flow as shown in Figure 4.10. This results in a switch point with an α body angle of about −57° when the tail angle is θ of 20° as shown in Figure 4.8(e). It should be noted, however, that the exact angle values of the switch points might depend on the position of the center of mass, which can only be estimated and not accurately measured.

These results are evidence of the major effect of the caudal fin on the hydrodynamically unstable body pathway of boxfish. In addition, the CFD simulation gives further insight into the distribution of pressure and net torque over the body, the caudal peduncle, and the caudal fin (closed and open) separately. This simulation confirms our experimental results. During large positive body angles α, the static equilibrium of hydrodynamic forces around the center of volume prevails over the body of the trunkfish, caudal peduncle and caudal fin compression forces induced on the caudal fin as shown in Figures 4.8(e), 4.8(f) and 4.9. The pressure exerted on the caudal fin exceeds the positive pressure on the head (see Figure 4.9). This induces the right orientation observed at large positive body angles α. When the model is at the switching point, the stress on the caudal fin and the caudal peduncle is equal compared to the head, resulting in a net torque of 0 N.m (i.e. neutral). When the model is placed at large negative body angles α, the pressure on the caudal fin is greater than on the head, which leads to steering to the left. The CFD results support the experimental results that the caudal fin is a major contributor to the orientation of the boxfish depending on the caudal fin condition and the tail angle θ as shown in Figure 4.10.

Results of this study indicate that boxfishes are (actively) controlling instability, rather than relying on the hydrodynamic stability inherent in the rigid structures of their body that will need to be corrected as the maneuver begins. The original idea that a boxfish shield might self-stabilize the flow (i.e. automatically re-align the body to the swimming direction) implied the need to increase the forces exerted by the fins to rotate the fish while swimming to negate the torque of self-stabilization on the shield. However, the entire shield destabilizes diffraction as shown in Figure 4.6. Hence, performing spin maneuvers would require less force from the fins or would result in faster rotations to introduce a certain force from the fins. The downside to the latter position is that recoil movements are more likely to occur while swimming, and can be very expensive. Previously detected vortices causing self-correcting forces on the shield will reduce these recoil motions. However, there is a trade-off between forceful costly recoil movements during routine swimming (when controlling instability) versus work required to perform the maneuvers (when correcting for stability), and the specific compromise will depend on the fitness consequences of the species. Perhaps the reason why boxfishes developed a locomotion system that controlled instability is related to their maneuverable lifestyle.

In conclusion, the results of this study quantitatively show that the effective change of caudal fin shape and direction plays

an important role in controlling steering torque in boxfish. The
caudal fin is both a track anchor and a rudder. These results
correlate with the swimming behavior of boxfish in their natural
environment. Although this new information is an important first
step in improving our understanding of controlling steering motions
in boxfishes, further study is needed to reveal how all components of
the boxfish thickness motor system work together, from a dynamic
perspective, during lateral storm flows and transformation.

4.3. Steering by Pectoral Fin in Swimming Robot

This section aims to develop a swimming robot with good steering
performance, in which the steering behavior is achieved by one degree
of freedom (1-DOF) represented by two concave-shaped pectoral fins.
The steering mechanism adopted here is based on the differential
drive principle. This principle is carried out by varying the right/left
fin velocities. Different radii have been achieved with four different
cases of velocities. The proposed design is validated theoretically via
the SOLIDWORKS® platform and proved practically in a physical
swimming pool.

4.3.1. *Steering Process*

To achieve the steering process of the swimming robot, a differential
drive mechanism of two-wheel mobile robot, is adopted for this
purpose. The pectoral fins are grouped on a common axis, and each
fin can separately be driven either forward or backward. The velocity
of each fin can be tuned in an independent way. For the swimming
robot to perform the turning motion, the robot must rotate about a
specific point. This point is located along its common left and right
fin axes and it is defined as the Instantaneous Center of Curvature
(ICC) as shown in Figure 4.11.

By varying the velocities of the two fins, the trajectories of the
swimming robot will change. Due to the velocity rate of rotation w_t
about the ICC must be the same for both fins, so left and right

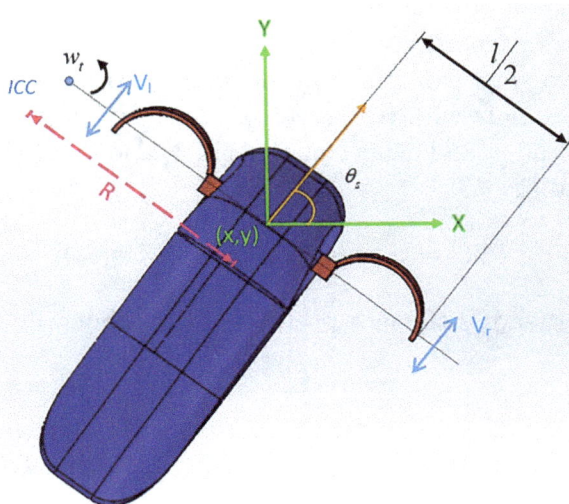

Figure 4.11. Differential drive kinematics for swimming robot.

velocities can be obtained as given:

$$V_r = w_t(R + 1/2) \tag{4.1}$$

$$V_l = w_t(R - 1/2) \tag{4.2}$$

where l can be defined as the distance that lies between the centers of the two fins, V_r, V_l are the right and left fin velocities, whereas R is the distance from the *ICC* calculated point to the midpoint between the fins.

At any moment, R and w_t can be solved as:

$$R = \frac{l}{2}\frac{V_l + V_r}{V_r - V_l} \tag{4.3}$$

$$w_t = \frac{V_r - V_l}{l} \tag{4.4}$$

The robot is initially set at a position (x, y), it is directed such that it makes an angle s with X-axis (refer to Figure 4.11), which assumes that the robot is centered at a point midway along the fin axel.

By varying v_l and v_r, the swimming robot can be moved to different positions and orientations. In order to obtain the turning radius of the robot, we are seeking for the change in position and orientation of (x, y) coordinate with respect to the time. By calculating the two velocities v_l and v_r and using equations (4.3) and (4.4), the ICC point can be found as:

$$ICC = [x - R\sin(\theta_s), y + R\cos(\theta_s)] \tag{4.5}$$

At time t and δt the new position of the robot will be:

$$\begin{bmatrix} \dot{x} \\ \dot{y} \\ \dot{\theta}_s \end{bmatrix} = \begin{bmatrix} \cos(w_t \delta t) & -\sin(w_t \delta t) & 0 \\ \sin(w_t \delta t) & \cos(w_t \delta t) & 0 \\ 0 & 0 & 1 \end{bmatrix}$$
$$\times \begin{bmatrix} x - ICC_x \\ y - ICC_y \\ \theta_s \end{bmatrix} + \begin{bmatrix} ICC_x \\ ICC_y \\ w_t \delta t \end{bmatrix} \tag{4.6}$$

4.3.2. *Simulation and Experimental Results*

The proposed design of the swimming robot is simulated by SOLID-WORKS® software and validated experimentally. The starting angle of rotation is set to 50°. The robot will oscillate in a 100° amplitude (i.e. from 50° to −50°). A computational domain of exact dimensions of the swimming pool (i.e. 1 m × 0.65 m × 0.65 m) is used throughout this experiment as in Figure 4.12. The flow type is set to have laminar and turbulent options. The static pressure of 101325 Pa at 293.2 K is used, and a local mesh of six levels of refinement cells are used throughout this simulation.

4.3.2.1. *Experimental setup*

The concave shape is utilized by the pectoral fin to produce a maximum thrust during the power stroke and minimum drag at recovery stroke. The model is tested with different power to recovery stroke ratios (i.e. RF of 2:1, 3:1, 4:1, and 5:1) of left fin and (1:1) of the right one.

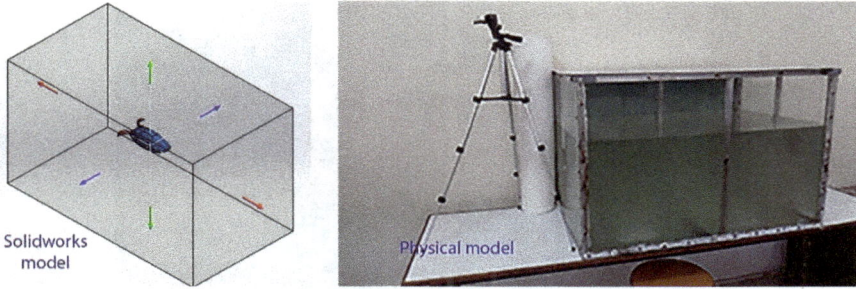

Figure 4.12. Physical and SOLIDWORKS® environments.

The robot motion is recorded by a Kodak camera with high resolution specifications at 30 frames per second (fps). The recording camera, is fixed on the top of the swimming pool at 50 cm distance from the robot to capture the motion of the robot. Three black-labels are fixed on the robot's body for motion tracking. Motion commands are sent to the controller via the HC-06 Bluetooth module, four 1.5 V AA batteries are used to supply the robot with the required energy as shown in Figure 4.13. Water density is assumed to be 1000 kg/m³.

Inertia coefficients and other parameters concerning the robot's body are the same as given in Chapter 3 (refer to Table 3.1).

4.3.2.2. *Experimental results*

For each experiment, the designed robot swam for approximately 20 sec until it reached a steady-state motion, after that, the steady-state data were recorded and extracted. Fin actuating is as shown in Figure 4.14. The red curve represents the velocity of the right fin, and it is the same for all four cases. The variations of the speed for the left fin for the following four cases are denoted by green, purple, yellow, blue curves, respectively. The right fin oscillates at $R_F = 1$, the power stroke phase starts from 0 sec to 0.33 sec, while power stroke starts from 0.33 sec to 0.66 sec, whereas the left fin is actuated in four different cases for different fin beat frequencies as follows:

Case 1: The velocity of the left fin at power stroke is twice the velocity of the right fin at recovery stroke $R_F = 2$, the power stroke

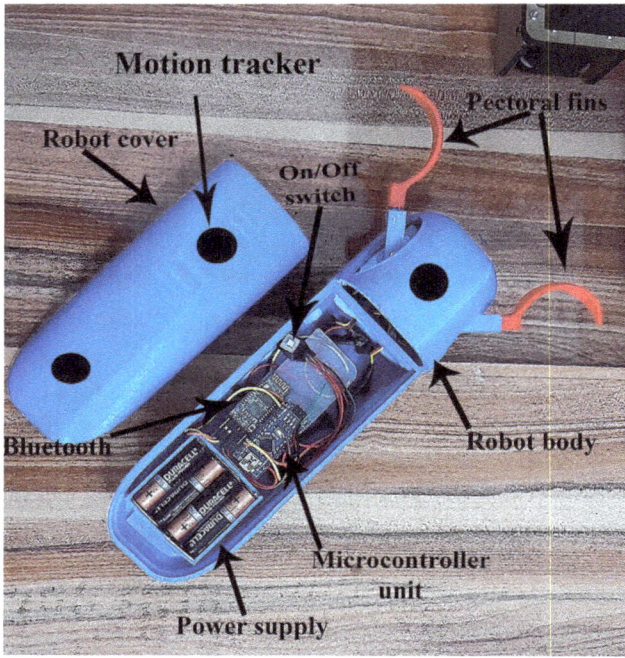

Figure 4.13. Physical model of the proposed robot.

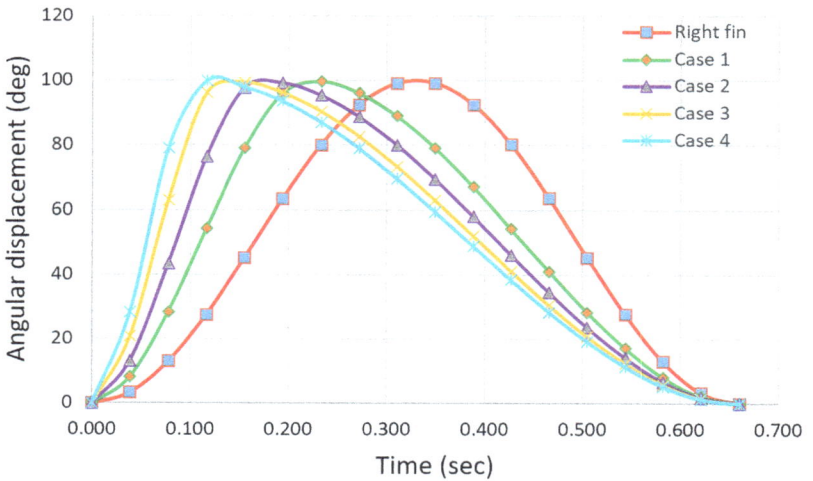

Figure 4.14. Fin actuation signals.

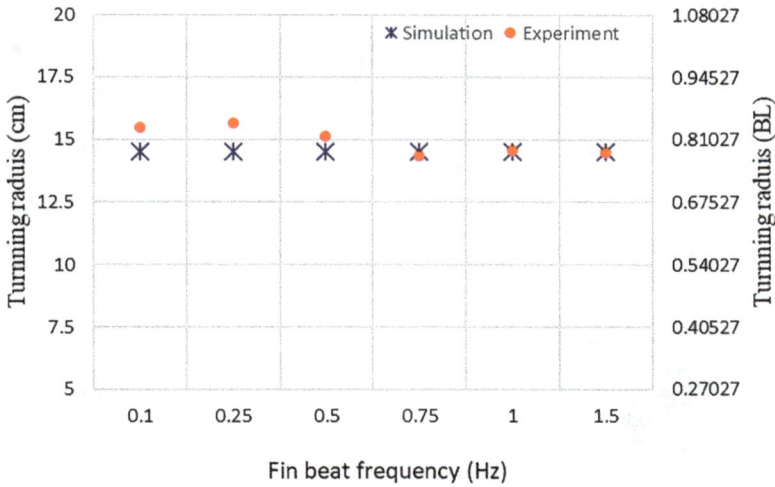

Figure 4.15. Turning radius in Case 1.

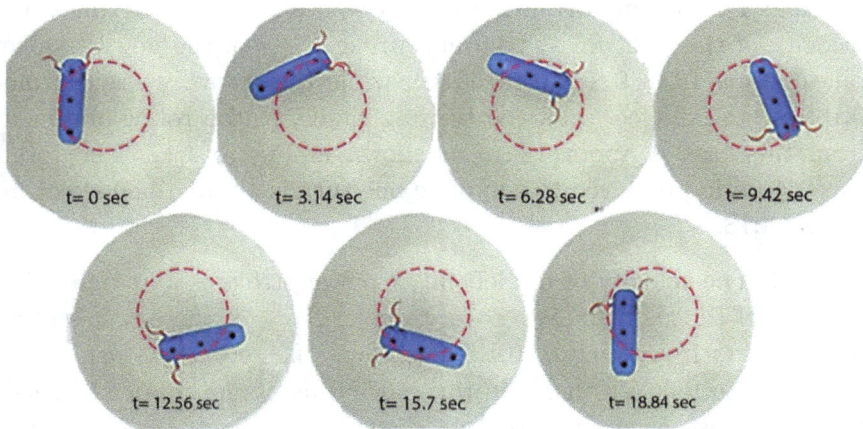

Figure 4.16. Turning period in Case 1.

phase starts from 0 sec to 0.22 sec, while the recovery stroke starts from 0.22 sec to 0.66 sec, then the turning radius of the robot obtained from this case is about 14 cm as shown in Figure 4.15. The total turning period required to complete one turn is about 18.84 sec as shown in Figure 4.16.

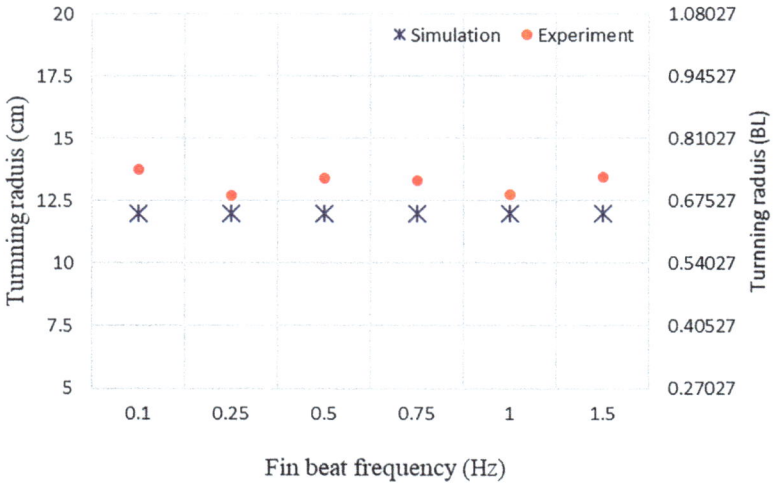

Figure 4.17. Turning radius in Case 2.

Case 2: The velocity of the left fin at power stroke is three times the velocity of the right fin at recovery stroke $R_F = 3$, the power stroke phase starts from 0 sec to 0.165 sec, while the recovery starts from 0.165 sec to 0.66 sec, then the turning radius of the robot obtained from this case is about 12 cm as shown in Figure 4.17. The total turning period required to complete one turn is about 16.80 sec as in Figure 4.18.

Case 3: The velocity of the left fin at power stroke is four times the velocity of the right fin at recovery stroke $R_F = 4$, the power stroke phase starts from 0 sec to 0.132 sec, while the recovery starts from 0.132 sec to 0.66 sec then the turning radius of the robot obtained from this case is about 10 cm as shown in Figure 4.19. The total turning period required to complete one turn is about 15.84 sec as shown in Figure 4.20.

Case 4: The velocity of the left fin at power stroke is twice the velocity of the right fin at recovery stroke $R_F = 5$, the power stroke phase starts from 0 sec to 0.11 sec, while the recovery starts from 0.11 sec to 0.66 sec, then the turning radius of the robot obtained from this case is about 8 cm as shown in Figure 4.21. The total turning

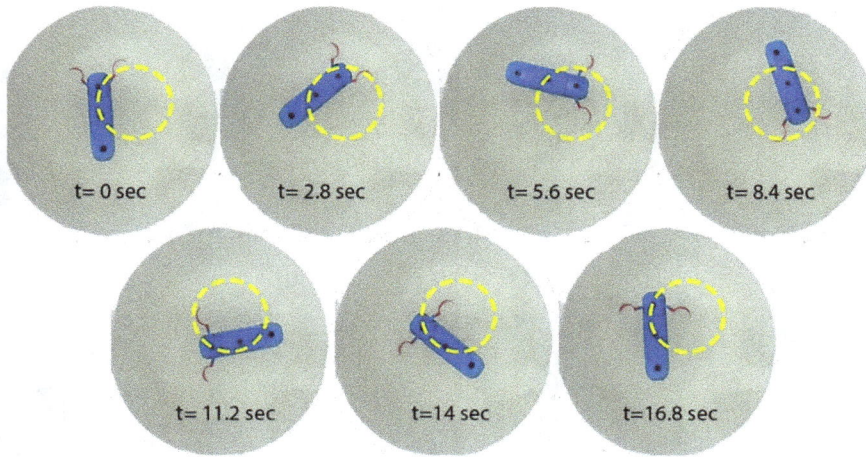

Figure 4.18. Turning period in Case 2.

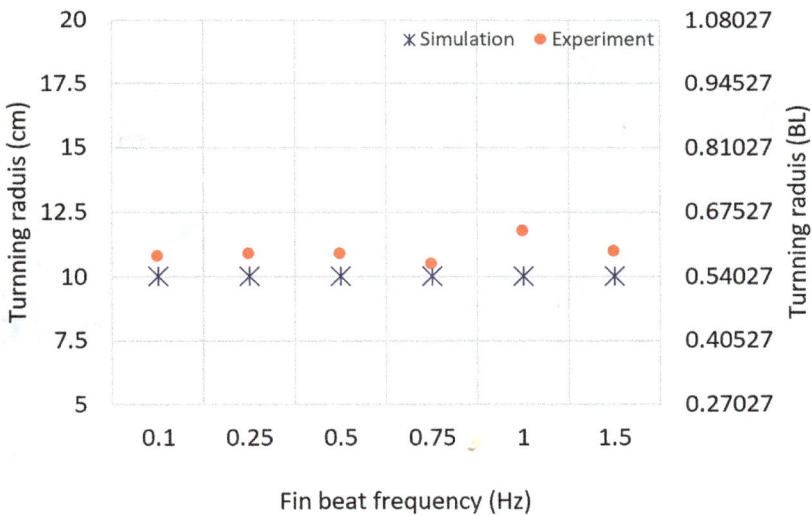

Figure 4.19. Turning radius in Case 3.

period required to complete one turn is about 14.82 sec as shown in Figure 4.22.

From the previous figures, it can be noticed that there is negative correlation relationship between the turning radius and the increase

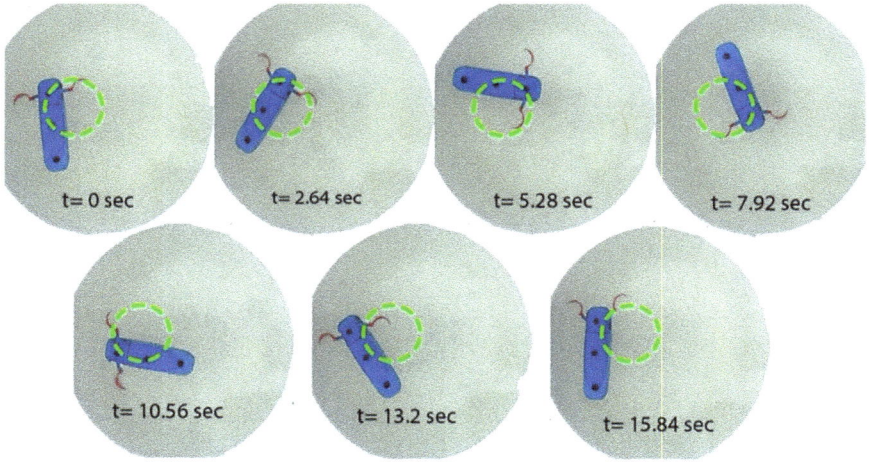

Figure 4.20. Turning period in Case 3.

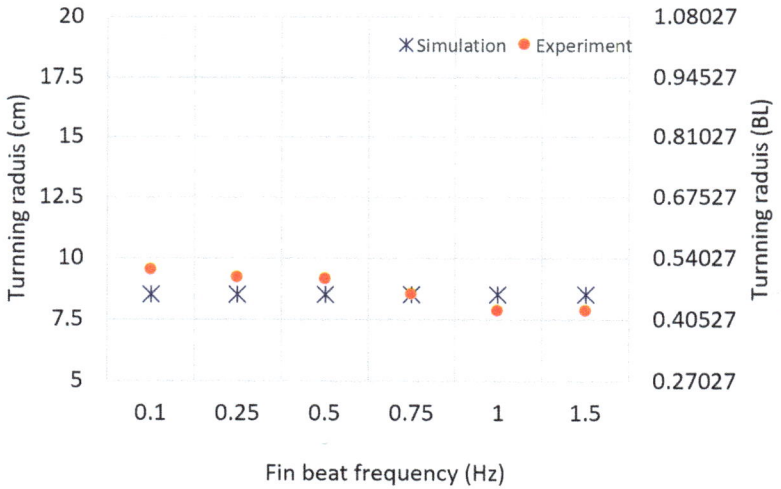

Figure 4.21. Turning radius in Case 4.

in fin beat frequencies. The results show that the minimum turning radius reaches about 0.40 of body length for a turning period of about 15 sec to complete one turn in Case 4. These results outperform the previous ones recorded in literature as shown in Table 4.2.

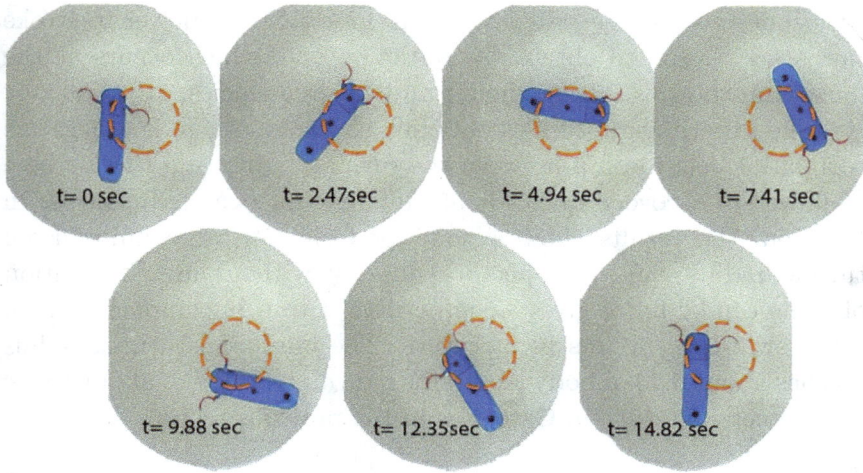

Figure 4.22. Turning period in Case 4.

Table 4.2. The swimming performance comparison of several searches.

References	Average turning radius
[26]	0.23–0.24 m (1.53–1.60 BL)
[35]	0.25–0.33 m (0.63–0.83 BL)
[112]	0.16–0.18 m (1.07–1.20 BL)
[113]	0.16–0.23 m (1.07–1.53 BL)
This work	0.08–0.095 m (0.40–0.50 BL)

4.3.3. *Conclusions*

A model of swimming robot actuated through a pair of concave pectoral fins has been presented. It is good to emphasize that the proposed model is not to fully replicate the pectoral fin morphology or perform exactly like the real fish. The goal is to provide a simple, low-cost, and high-performance robotic fish with steering capabilities, utilizing the principle of a differential drive of a two-wheel mobile robot to perform steering tasks of the swimming robot.

The velocities of the two pectoral fins are varied from each other, one fin is fixed within the same power to recovery ratio while the other one was with several ratios, and four different power to stroke ratios are chosen to obtain the optimum one, which occurs at the maximum ratio within maximum fin beat frequencies.

The experimental results obtained show that there is a negative correlation relationship between the turning radius and the increase of power to recovery ratio for all different fin beat frequencies. In addition, the results show a good performance when minimizing turning radius while using pectoral fins only without any intervention of using caudal fin; it gives a competitive output in comparison with other swimming robots in literature. The minimum turning radius reaches about 0.40 of body length for a turning period of about 15 sec to complete one turn in Case 4 as in the previous section.

References

Anderson, J. M. and Chhabra, N. K. (2002). Maneuvering and stability performance of a robotic tuna, *Integrative and Comparative Biology*, 42(1), pp. 118–126.

Bandyopadhyay, P. R. (2004). Guest editorial: Biology-inspired science and technology for autonomous underwater vehicles, *IEEE Journal of Oceanic Engineering*, 29(3), pp. 542–546.

Bartol, I. K., Gharib, M., Webb, P. W., Weihs, D. and Gordon, M. S. (2005) Body-induced vortical flows: A common mechanism for self-corrective trimming control in boxfishes, *J. Exp. Biol.*, 208, pp. 327–344. (doi:10.1242/jeb.01356)

Bartol, I. K., Gharib, M., Weihs, D., Webb, P. W., Hove, J. R. and Gordon, M. S. (2003). Hydrodynamic stability of swimming in ostraciid fishes: Role of the carapace in the smooth trunkfish Lactophrys triqueter (Teleostei: Ostraciidae), *J. Exp. Biol.*, 206, pp. 725–744. (doi:10.1242/jeb.00137)

Behbahani, S. B., Wang, J. and Tan, X. (2013). A dynamic model for robotic fish with flexible pectoral fins, *IEEE/ASME International Conference on Advanced Intelligent Mechatronics IEEE*, pp. 1552–1557.

Blake, R. W. (1977). On ostraciiform locomotion, *J. Mar. Biol. Assoc. UK*, 57, pp. 1047–1055. (doi:10.1017/S0025315400026114)

Blake, R. W., Chatters, L. M. and Domenici, P. (1995). The turning radius of yellowfin tuna (Thunnus albacores) in unsteady swimming manoeuvres, *Journal of Fish Biol.*, 46, pp. 536–538.

Boute, P. G., Van, W. S. and Stamhuis, E. J. (2020). Modulating yaw with an unstable rigid body and a course-stabilizing or steering caudal fin in the yellow boxfish (Ostracion cubicus)., *R. Soc. Open Sci.*, 7, pp. 200129. http://dx.doi.org/10.1098/rsos.200129.

Breder, C. M. (1926). The locomotion of fishes, *Zoologica*, 4, pp. 159–297.

Cheneval, O., Blake, R. W., Trites, A. W. and Chan, K. H. S. (2007). Turning maneuvers in Steller sea lions (Eumatopias jubatus), *Marine Mammal Science*, 23, pp. 94–109.

Comeau, S., Edmunds, P. J., Lantz, C. A. and Carpenter, R. C. (2014). Water flow modulates the response of coral reef communities to ocean acidification, *Sci. Rep.*, 4, 6681. (doi:10.1038/srep06681)

Domenici, P. (2001). The scaling of locomotor performance in predator-prey encounters: From fish to killer whales, *Comparative Biochemistry and Physiology Part A*, 131, pp. 169–182.

Domenici, P. and Blake, R. W. (1997). The kinematics and performance of fish fast-start swimming, *Journal of Experimental Biology*, 200, pp. 1165–1178.

Domenici, P., Standen, E. M. and Levine, R. P. (2004). Escape manoeuvres in the spiny dogfish (Squalus acanthias), *Journal of Experimental Biology*, 207, pp. 2339–2349.

Dudek, G. and Jenkin, M. R. M. (2000). Computational principles of mobile robotics, *Engineering, Computer Science*, Cambridge University Press.

Fish, F. E. (2002). Balancing requirements for stability and maneuverability in cetaceans, *Integrative and Comparative Biology*, 42, pp. 85–93.

Fish, F. E. (2004). Structure and mechanics of nonpiscine control surfaces, *IEEE Journal of Oceanic Engineering*, 28(3), pp. 605–621.

Fish, F. E. and Lauder, G. V. (2017). Control surfaces of aquatic vertebrates in relation to swimming modes, *Journal of Experimental Biology*, 220, pp. 4351–4363.

Fish, F. E. and Lauder, G. V. (2017). Control surfaces of aquatic vertebrates . in relation to swimming modes, *Journal of Experimental Biology*, 220, pp. 4351–4363.

Fish, F. E. and Nicastro, A. J. (2003). Aquatic turning performance by the whirligig beetle: constraints on maneuverability by a rigid biological system, *Journal of Experimental Biology*, 206(10), pp. 1649–1656.

Fish, F. E. and Nicastro, A. J. (2003). Aquatic turning performance by the whirligig beetle: constraints on maneuverability by a rigid biological system, *Journal of Experimental Biology*, 206(10), pp. 1649–1656.

Fish, F. E., Hurley, J. and Costa, D. P. (2003a). Maneuverability by the sea lion, Zalophus californianus: Turning performance of an unstable body design, *Journal of Experimental Biology*, 206, pp. 667–674.

Frey, E. and Salisbury, S. W. (2001). *The kinematics of aquatic locomotion in Osteolaemus tetraspis Cope, Crocodilian Biology and Evolution* ed. G. C. Grigg, F. Seebacher and C. E. Franklin (Chipping Norton, Australia: Surrey Beatty), pp. 165–179.

Geder, J. D., Ramamurti, R., Palmisano, J., Pruessner, M., Ratna, B. and Sandberg, W. C. (2011). Four-fin bio-inspired UUV: Modeling and Control Solutions, (No. IMECE2011-64005) (Washington DC: Naval Research Lab).

Geurten, B. R., Niesterok, B., Dehnhardt, G. and Hanke, F. D. (2017). Saccadic movement strategy in a semiaquatic species—the harbour seal (Phoca vitulina), *Journal of Experimental Biology*, 220, pp. 1503–1508.

Gordon, M. S., Hove, J. R., Webb, P. W. and Weihs, D. (2000). Boxfishes as unusually well-controlled autonomous underwater vehicles, *Physiol. Biochem. Zool.*, 73, pp. 663–671. (doi:10.1086/318098)

Gordon, M. S., Plaut, I. and Kim, D. (1996). How puffers (Teleostei: Tetraodontidae) swim, *J. Fish Biol.*, 49, pp. 319–328. (doi:10.1111/j.1095-8649.1996. tb00026.x)

Gray, J. (1933). Directional control of fish movement, *Proc. R. Soc. Lond. B*, 113, pp. 115–125. (doi:10.1098/rspb.1933.0035)

Gray, J. (1968). *Animal Locomotion*, London, UK: Weidenfeld and Nicholson.

Heatwole, S. J. and Fulton, C. J. (2013). Behavioural flexibility in reef fishes responding to a rapidly changing wave environment, *Mar. Biol.*, 160, pp. 677–689. (doi:10.1007/s00227-012-2123-2)

Helmer, D., Geurten, B. R., Dehnhardt, G. and Hanke, F. D. (2017). Saccadic movement strategy in common cuttlefish (Sepia officinalis), *Frontiers in Physiology*, 7, pp. 660.

Hench, J. L. and Rosman, J. H. (2013). Observations of spatial flow patterns at the coral colony scale on a shallow reef flat, *J. Geophys. Res. Oceans*, 118, pp. 1142–1156. (doi:10.1002/jgrc.20105)

Hirata, K., Takimoto, T. and Tamura, K. (2000). Study on turning performance of a fish robot, *First International Symposium on Aqua Bio-Mechanisms*, pp. 287–292.

Hove, J. R., O'Bryan, L. M., Gordon, M. S., Webb, P. W. and Weihs, D. (2001). Boxfishes (Teleostei: Ostraciidae) as a model system for fishes swimming with many fins: Kinematics, *J. Exp. Biol.*, 204, pp. 1459–1471.

Howland, H. C. (1974). Optimal strategies for predator avoidance: The relative importance of speed and manoeuvrability, *Journal of Theoretical Biology*, 47, pp. 333–350.

Hui, C. A. (1985). Maneuverability of the Humboldt penguin (Spheniscus humboldti) during swimming, *Canadian Journal of Zoology*, 63, pp. 2165–2167.

Jastrebsky, R. A., Bartol, I. K. and Krueger, P. S. (2016). Turning performance in squid and cuttlefish: Unique dual-mode, muscular hydrostatic systems, *Journal of Experimental Biology*, 219, pp. 1317–1326.

Jastrebsky, R. A., Bartol, I. K. and Krueger, P. S. (2017). Turning performance of brief squid Lolliguncula brevis during attacks on shrimp and fish, *Journal of Experimental Biology*, 220, pp. 908–919.

Johansen, J. L. (2014). Quantifying water flow within aquatic ecosystems using load cell sensors: A profile of currents experienced by coral reef organisms around Lizard Island, Great Barrier Reef, Australia, *PLoS ONE* 9, e83240. (doi:10.1371/journal.pone.0083240)

Kajiura, S. M., Forni, J. B. and Summers, A. P. (2003). Maneuvering in juvenile carcharhinid and sphyrnid sharks: The role of the hammerhead shark cephalofoil, *Zoology*, 106, pp. 19–28.

Kumph, J. M. (2000). *Maneuvering of a Robotic Pike*, MS Thesis (Cambridge, MA: Massachusetts Institute of Technology).

Li, Z., Du, R., Zhang, Y. and Li, H. (2013). Robot fish with novel wire-driven continuum flapping propulsion, *Applied Mechanics and Materials*, pp. 300–301, pp. 510–514.

Liang, J., Wang, T. and Wen, L. (2011). Development of a two-joint robotic fish for real-world exploration, *Journal of Field Robotics*, 28(1), pp. 70–79.

Mayerl, C. J., Youngblood, J. P., Rivera, G., Vance, J. T. and Blob, R. W. (2018). Variation in morphology and kinematics underlies variation in swimming stability and turning performance in freshwater turtles, *Integrative Organismal Biology*, 1(1), p. oby001.

Menozzi, A., Leinhos, H. A., Beal, D. N. and Bandyopadhyay, P. R. (2008). Open-loop control of a multifin biorobotic rigid underwater vehicle, *IEEE Journal of Oceanic Engineering*, 33(2), pp. 59–68.

Miller, D. (1991). *Submarines of the World* (New York: Orion Books).

Myers, R. F. (1999). *Micronesian Reef Fishes: A Comprehensive Guide to the Coral Reef Fishes of Micronesia* (3rd revised and expanded edition), Barrigada, Guam: Coral Graphics.

Naser, F. A. and Rashid, M. T. (2019). Design, modeling, and experimental validation of a concave-shape pectoral fin of labriform-mode swimming robot, *Engineering Reports*, 1(5), pp. 1–17.

Naser, F. A. and Rashid, M. T. (2020). Effect of Reynold number and angle of attack on the hydrodynamic forces generated from a bionic concave pectoral fins, *IOP Conf. Ser.: Mater. Sci. Eng.*, 745, pp. 1–13.

Naser, F. A. and Rashid, M. T. (2020). The influence of concave pectoral fin morphology in the performance of labriform swimming robot, *Iraqi Journal for Electrical and Electronic Engineering*, 16(1), pp. 54–61.

Naser, F. A. and Rashid, M. T. (2021). Design and realization of labriform mode swimming robot based on concave pectoral fins, *Journal of Applied Nonlinear Dynamics*, 10(4), pp. 691–710.

Naser, F. A. and Rashid, M. T. (2021). Enhancement of labriform swimming robot performance based on morphological properties of pectoral fins, *J Control Autom Electr Syst*, 32, pp. 927–941.

Naser, F. A. and Rashid, M. T. (2021). Implementation of steering process for labriform swimming robot based on differential drive principle, *Journal of Applied Nonlinear Dynamics*, 10(4), pp. 737–753.

Naser, F. A. and Rashid, M. T. (2021). Labriform swimming robot with steering and diving capabilities, *Journal of Intelligent & Robotic Systems*, 103(14), pp. 1–19.

Parson, J., Fish, F. E. and Nicastro, A. J. (2011). Turning performance in batoid rays: Limitations of a rigid body, *Journal of Experimental Marine Biology and Ecology*, 402, pp. 12–18.

Plaut, I. and Chen, T. (2003). How small puffers (Teleostei: Tetraodontidae) swim, *Ichthyol. Res.*, 50, pp. 149–153. (doi:10.1007/s10228-002-0153-3)

Pollard, B. and Tallapragada, P. (2019). Passive appendages improve the maneuverability of fishlike robots, *IEEE/ASME Transactions on Mechatronics*, 24(4), pp. 1586–1596.

Porter, M. E., Roque, C. M. and Long Jr. J. H. (2011). Swimming fundamentals: Turning performance of leopard sharks (Triakis semifasciata) is predicted by body shape and postural reconfiguration, *Zoology*, 114(6), pp. 348–359.

Rashid, M. T. and Rashid, A. T. (2016). Design and implementation of swimming robot based on labriform model, *Al-Sadeq International Conference on Multidisciplinary in IT and Communication Science and Applications (AIC-MITCSA)*, pp. 1–6.

Rashid, M. T., Naser, F. A. and Mjily, A. H. (2020). Autonomous micro-robot like sperm based on piezoelectric actuator, *International Conference on Electrical, Communication, and Computer Engineering (ICECCE)*, pp. 1–6.

Rivera, G., Rivera, A. R. V., Dougherty, E. E. and Blob, R. W. (2006). Aquatic turning performance of painted turtles (Chrysemys picta) and functional consequences of a rigid body design, *Journal of Experimental Biology*, 209, pp. 4203–4213.

Schneider, C. A., Rasband, W. S. and Eliceiri, K. W. (2012). NIH Image to ImageJ: 25 years of image analysis, *Nat. Methods*, 9, pp. 671–675. (doi:10.1038/nmeth.2089)

Stanway, J. (2008). The turtle and the robot, *Oceanus*, 47, pp. 22–25.

Su, Z., Yu, J., Tan, M. and Zhang, J. (2013). Implementing flexible and fast turning maneuvers of a multijoint robotic fish, *IEEE/ASME Transactions on Mechatronics*, 19(1), pp. 329–338.

Triantafyllou, M. S. (2017). Tuna fin hydraulics inspire aquatic robotics, *Science*, 357, pp. 251–252.

Van, W. S., Van, M. K., Marcroft, T. A., Alfaro, M. E. and Stamhuis, E. J. (2015). Boxfish swimming paradox resolved: Forces by the flow of water around the body promote manoeuvrability, *J. R. Soc. Interface*, 12, 20141146. (doi:10.1098/rsif.2014.1146)

Videler, J. J. (1993). *Fish Swimming*, London, UK: Chapman & Hall.

Walker, J. A. (2000). Does a rigid body limit maneuverability? *Journal of Experimental Biology*, 203, pp. 3391–3396.

Walker, J. A. (2000). Does a rigid body limit maneuverability? *J. Exp. Biol.*, 203, pp. 3391–3396.

Walker, J. A. (2004). Kinematics and performance of maneuvering control surfaces in teleosts fishes, *IEEE Journal of Oceanic Engineering*, 29, pp. 572–584.

Webb, P. W. (1976). The effect of size on the fast-start performance of rainbow trout. Salmo gairdneri, and a consideration of piscivorous predator-prey interactions, *Journal of Experimental Biology*, 65, pp. 157–177.

Webb, P. W. (1983). Speed, acceleration and manoeuverability of two teleost fishes, *Journal of Experimental Biology*, 102, pp. 115–122.

Webb, P. W. (1994). The biology of fish swimming, *In Mechanics and Physiology of Animal Swimming* (eds L Maddock, Q Bone, JMV Rayner), pp. 45–62. Cambridge: Cambridge University Press.

Webb, P. W. (2004). Maneuverability-general issues, *IEEE Journal of Oceanic Engineering*, 29, pp. 547–555.

Webb, P. W. (2011). Buoyancy, locomotion, and movement in fishes: Maneuverability, *In Encyclopedia of Fish Physiology: From Genome to Environment* (ed. AP Farrell), pp. 575–580. London, UK: Academic Press.

Weihs, D. (1972). A hydrodynamical analysis of fish turning manoeuvres, *Proc. R. Soc. Lond. B*, 182, pp. 59–72. (doi:10.1098/rspb.1972.0066)

Wu, Z., Yu, J., Yuan, J. and Tan, M. (2019). Towards a gliding robotic dolphin: Design, modeling, and experiments, *IEEE/ASME Transactions on Mechatronics*, 24(1), pp. 260–270.

Ye, X., Su, Y. and Guo, S. (2007). A centimeter-scale autonomous robotic fish actuated by IPMC actuator, *IEEE International Conference on Robotics and Biomimetics (ROBIO) IEEE*, pp. 262–267.

Yu, J., Liu, L., Wang, L., Tan, M. and Xu, D. (2008). Turning control of a multilink biomimetic robotic fish, *IEEE Transactions on Robotics*, 24(1), pp. 201–206.

Yu, J., Su, Z., Wang, M., Tan, M. and Zhang, J. (2012). Control of yaw and pitch maneuvers of a multilink dolphin robot, *IEEE Transactions on Robotics*, 28(2), pp. 318–329.

Yu, J., Zhang, C. and Liu, L. (2016a). Design and control of a single-motor-actuated robotic fish capable of fast swimming and maneuverability, *IEEE/ASME Transactions on Mechatronics*, 21(3), pp. 1711–1719.

Zhang, F., Zhang, F. and Tan, X. (2014). Tail-enabled spiraling maneuver for gliding robotic fish, *ASME Journal of Dynamic Systems, Measurement and Control*, 136(4), 041028.

Zhu, Q. and Shoele, K. (2008). Propulsion performance of skeleton-strengthened fin, *J. Exp. Biol.*, 211, pp. 2087–2100. (doi:10.1242/jeb. 016279)

Chapter 5

DIVING PROCESS
OF SWIMMING ROBOT

5.1. Introduction

For the diving process, the propulsion should be in the direction of upward and downward sliding of the swimming robot in the water. This process is achieved by controlling the swimming robot buoyancy by using the self-actuating mechanism. Therefore, vertical motion will be generated when changing the swimming robot buoyancy, which converts the 2 DOF motion of the swimming robot to 3 DOF motion. The upward and downward movements of the swimming robot follow a sawtooth pattern. They make use of hydrodynamic wings to convert vertical motion into horizontal motion, moving forward with very low power consumption.

In order to describe the concept of diving in water, there are three types of diving methods: electric diving system, which is a battery-powered pump for changing buoyancy, a thermal diving system that utilizes water's temperature gradients as an energy source, and hybrid diving system, which uses a battery-powered pump by utilizing water temperature gradient. A few years ago, underwater vehicles were used widely for oceanographic research. It can navigate through water by changing the ballast water amount in the buoyancy system periodically. Many researchers adopted the dynamic models of underwater vehicles with the mass slider motion system.

Generally, swimming robots have high maneuverability, for example, in terms of swimming turning radius as compared with autonomous unmanned vehicles (AUV), but they require a fixed actuation for swimming, they cannot operate for a long period compared with AUV. The concept of diving in a swimming robot was developed by combining the desirable characteristics of an AUV with a swimming robot in one design. However, integrating fin-actuation mechanisms, with a center of gravity controller (CoG) principle is a key to the challenge.

After providing the complete dynamics and motion analysis for the pectoral fins (in Chapter 2), the body achieves forward swimming (in Chapter 3), steering (in Chapter 4). This chapter presents the design and modeling of the swimming robot that can achieve the diving process based on a sliding mass mechanism, while the diving systems that will be introduced in this chapter overcome the size challenges in the swimming robots, in which two systems will be described. The propulsion of the first swimming robot depends on Carangiform mode in swimming, and the second one depends on the principle of Labriform mechanism mode, while the diving system of these two models depends on the center of gravity controller (CoG) by using a sliding mass system.

5.2. Diving System of Carangiform Swimming Robot

In this section, the mechatronic design and fabrication of a prototype Carangiform (i-RoF) swimming robot with a two-link propulsion tail mechanism will be presented. For the design procedures, a multilink biomimetic approach is adapted, which uses the physical properties of real carp as its size and structure. The appropriate body rate is determined by the swimming patterns and oscillations of the caudal fin. The prototype consists of three main parts: a front rigid body, a two-link thruster tail mechanism, and a flexible tail fin. Prototype parts are implemented by 3D printing technology. A biomimetic motion control structure based on a central pattern generator (CPG) is proposed to simulate the vigorous, fish-like swimming gait. The designed unidirectional sequential CPG network is inspired by the lamprey neural spinal cord and generates stable rhythmic oscillatory patterns. In addition, the center of gravity (CoG) control mechanism is designed and placed in the front rigid body to ensure 3D swimming ability. With the help of this design, the characteristics of the swimming robot with swimming forward, turn, buoyancy, diving, and self-propelled movements in the experimental pool are implemented.

The biomimetic design and fabrication of an intelligent swimming robot (i-RoF) prototype for real-world exploration and survey tasks are proposed, these important tasks require high swimming ability and direct swimming speed. Therefore, important key issues related to 3D swimming capabilities and biomimetic design are emphasized and detailed in this study. A solid torpedo-shaped front body is designed to contain the electronics, sensors, and CoG control mechanism. The CoG control mechanism successfully provides buoyancy and diving movement capabilities. Motion control is adapted based on a central pattern generator (CPG) to generate strong, smooth, and regular oscillating swimming patterns. The proposed CPG model is inspired by the lamprey spinal cord to ensure the biomimetic design and intelligent control. In this regard, the proposed swimming robot prototype is biomimetic in terms of design and control. Swimming robot performance analyzes are performed in an experimental pool environment to verify the robot's high swimming capabilities.

5.2.1. *Design Procedure of the Swimming Robot*

The design procedures, content, and implementation of the swimming robot (i-RoF) which is manufactured at Firat University, are shown. The prototype, inspired by the Carangiform movement, has two key features: a fish-like swimming ability with a new propulsive tail mechanism design, 3D autonomous movement with surrounding sensors, and a smart controller.

The following eight basic characteristics are taken into account for the biomimetically designed robotic fish model:

(1) The Carangiform-inspired swimming robot features a bio-mimic design and can swim by mimicking the swimming motion of a real carp.

(2) The torpedo-shaped body of the robotic fish is designed to reduce hydrodynamic friction forces and the outer surface of each component is coated with materials that reduce vortex effects.

(3) The components of multiple parts of the robot are designed as modular units. Thus, the necessary interventions are relatively easy at the time of assembly and breakdown.

(4) All mechanical components are provided to be water-resistant. Thus, experimental studies of the prototype can be easily performed in water.

(5) CoG is located in the same direction as the center of the buoyancy and slightly lower. Thus, the effect of the robotic fish roll in the water stabilizes at about 0 degrees.

(6) The robotic fish has 3D mobility and the ability to stand on the surface of the water without applying thrust on the robot. It is provided by ensuring that the density of the robot is very close to $1000\,\text{g/m}^3$, which is the density of water.

(7) The robotic fish has an intelligent biomimetic structure and motion control that can perform complex control algorithms using CPG and Fuzzy Logic approaches.

(8) The robotic fish can swim independently by avoiding obstacles in 3D space.

5.2.2. *Mechanical Design of the Swimming Robot*

The swimming robot consists of five basic components including a rigid front main body, a servo-driven two-link tail mechanism, and a controller that performs the CPG model, a front sight unit, and a flexible tail fin. The tail links, which are driven by high-power servomotors and a single flexible tailfin, mounted on the bracket are designed to generate a moving body wave of the swimming robot. These links connected in the form of a chain structure produce the necessary thrust for the swimming movements.

Dimensions and properties of all parts are determined taking into account the body measurements of a true Carangiform fish. To analyze the two-link chain structure of a fish model, three points of verification of junction lengths are selected on the forefoot and swim

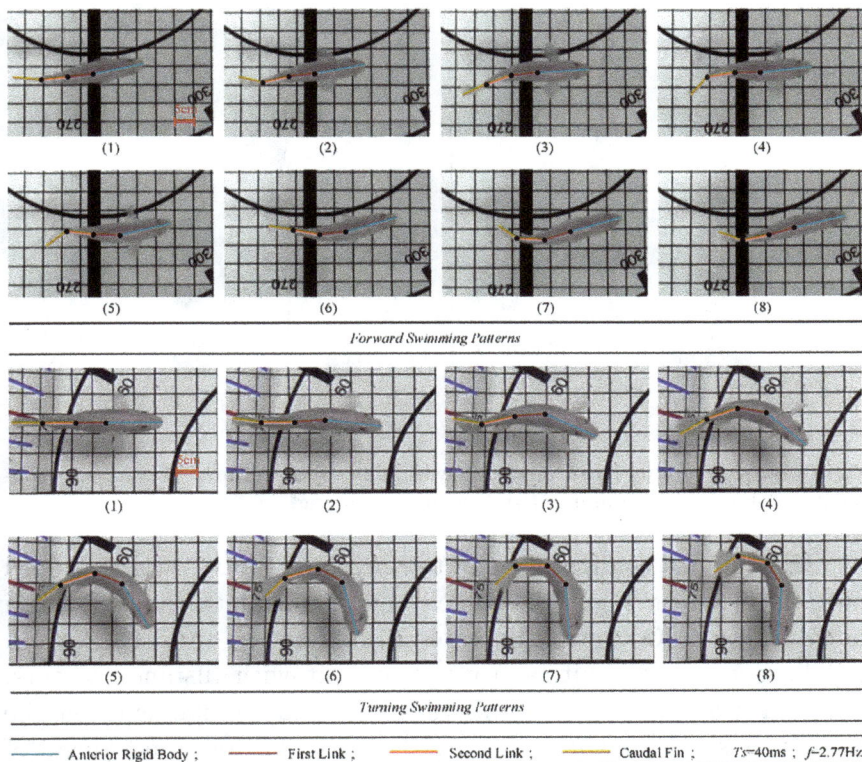

Figure 5.1. Forward and turning swimming patterns of a Carangiform carp fish for one period.

patterns of real fish as shown in Figure 5.1. The fact that robotic fish have as many joints as possible allows the movements of real fish to be more easily imitated. However, with the proposed tail mechanism, a simpler and more practical design can be achieved. Link lengths are determined by analyzing 50 swimming styles for the front and swimming pool positions. The junction lengths obtained from three basic components of real fish were adapted to the prototype model. The concept design of a swimming robot in a SimMechanics environment for forwarding swimming and turning patterns is shown in Figure 5.2.

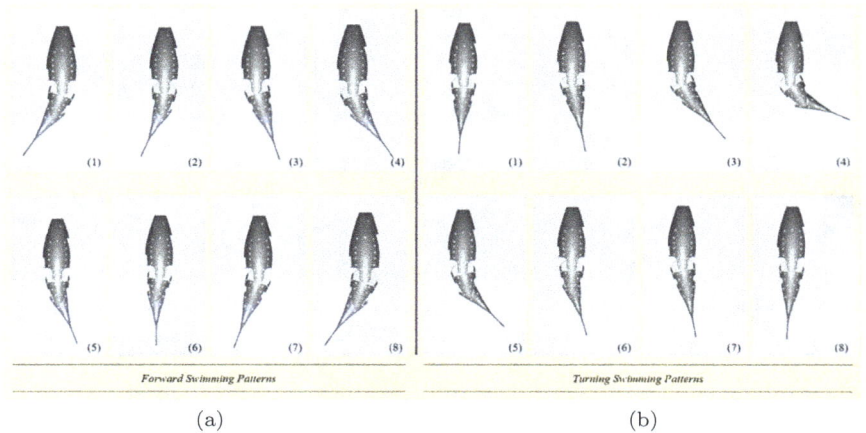

Figure 5.2. Swimming patterns of the swimming robot in SimMechanics environment: (a) Forward swimming. (b) Turning swimming.

The detailed configuration of the swimming robot with standard parts is shown in Figure 5.3. It can be seen that there is the main body with a CoG control mechanism and battery, two servo-driven tail fins, a control unit, a front sight unit with distance sensors, and a flexible tail fin. The distribution of all components on the prototype is carefully achieved to balance the lift force and weight of the robotic fish.

All components of the robotic fish are designed with proportional dimensions and the 3D solid model of each component is generated in the SolidWorks environment. At the production stage, the STL formats for the prepared solid models are transferred to the Voxelizer software. These files are converted to voxel format in the Voxelizer environment and layer, printing, and support settings are configured.

Every part of the prototype that is designed and 3D modeled using SolidWorks environment is produced using PLA in 3D printing technology. The flexible caudal fin is also produced using a silicone mold. All parts are covered with epoxy resin to prevent possible leakage from micro pores formed in the production process. After the assembly phase, the outer surface is covered with a synthetic coating to prevent leaks that may be caused by capillary cracks during assembly.

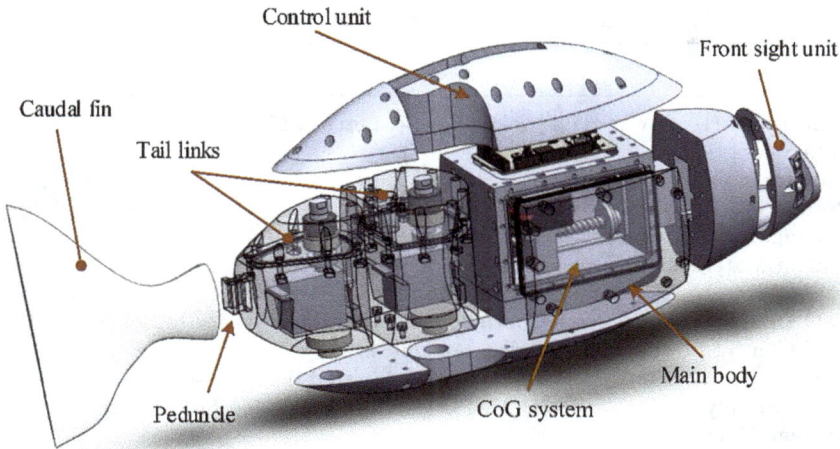

Figure 5.3. The detailed mechanical configuration of the swimming robot.

The CoG control mechanism is designed to perform the 3D movement in the water. The developed swimming robot can swim autonomously in the water and can perform the desired behavior using the commands sent by the user. Bluetooth communication is adapted for the wireless communication system between the swimming robot and the user's computer, and the wireless communication distance is about 10 m. In addition, the programming and charging connections of the prototype are implemented using the USB port. Swimming robot must be able to perceive static and/or dynamic obstacles in the environment as they move through the water. For this reason, three sharp infrared distance sensors were placed in the left, right, and front of the swimming robot. The Prop Shield 10-DoF IMU motion sensor is available for simultaneous angular and linear positioning, speed, acceleration, altitude, and temperature data while swimming. 7.4 V 1350 mAh Lithium Polymer (Li-Po) rechargeable battery was used to supply power to servomotors and electronic control system. Detailed technical characteristics of the developed swimming robot are shown in Table 5.1, and a photograph of the developed swimming robot is shown in Figure 5.4. The swimming robot is about 500 mm long, 76 mm wide, and 215 mm high, while its mass is about 3.1 kg.

Table 5.1. Technical characteristics of the swimming robot.

Items	Characteristics
Dimensions (L × W × H)	~500 mm × 76 mm × 215 mm
Total Mass	~3.1 kg
Number of Joints	2
Production Technique	3D-Printing Technology, Poly Lactic Acid (PLA)
Driving System	Servo motor (29 kg · cm, 7.4 V)
Sealing Materials	O-ring, Grease oil, Epoxy resin, Sealing paste Fluid seal, Rubber seal, Felt, Synthetic spray paint
CoG Driving System	Continuous rotation servo motor (12 kg · cm, 7.4 V)
CoG Control Mechanism	Ball screw-nut, lead
Sensors	10-DoF IMU, infrared distance sensors,
Power Supply	7.4 V, 1350 mAh rechargeable Li-Po battery
Average Operation Time	~30 min
Swimming Ability	Autonomous, user-controlled
Communication	Bluetooth, USB
Motion Ability	Three-dimensional
Operation Depth	Tested in maximum ~3 m

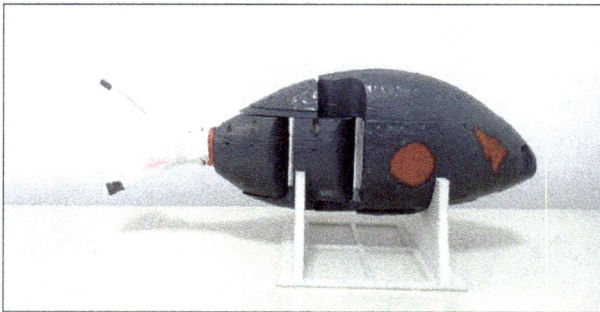

Figure 5.4. Developed swimming robot.

5.2.3. *Propulsive Two-Link Tail Mechanism*

The thrust necessary for the movement of the swimming robot is provided by a designed two-link tail mechanism. Each link consists of three basic components: a Hitec D954SW RC servomotor, a servomotor cover and an upper housing. The mechanical configuration of the designed second linkage is given in Figure 5.5. Each servomotor is mounted on the motor housing and an aluminum 25 T upper shaft is

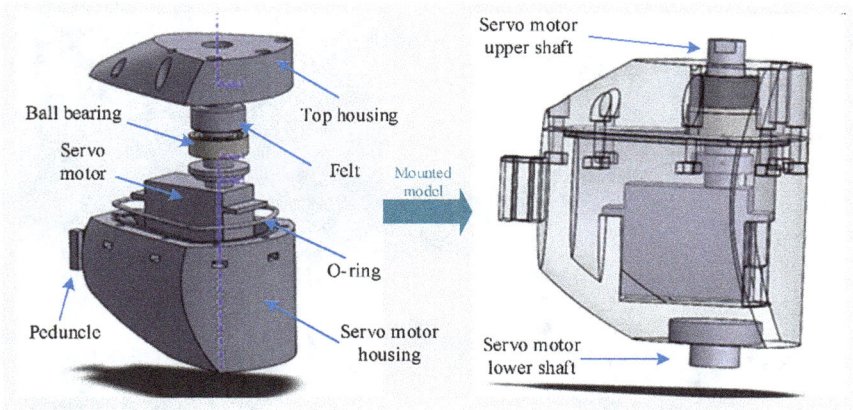

Figure 5.5. Detailed mechanical configuration of the second tail link: unmounted and mounted models.

installed on the motor gear shaft to transmit the rotational motion. A sealing felt and a single row fixed ball bearing are installed on the upper shaft seat so that the upper motor can perform rotational motion in the direction of the vertical axis and can be locked. Grease oil is used between the shaft and the ball bearing to increase the resistance of the upper shaft to water as it rotates. In addition, O-ring channels are designed and supplied with grease oil into these channels. The tail connection surfaces of the parts are covered with a liquid seal to increase water resistance. The bolt housing used for assembly is located at the top of the housing and the nuts used to compress the bolt are located on the motor housing. At the same time, the lower shaft of the servomotor is located in the vertical direction of the motor shaft and at the bottom of the motor bed to perform the free rotation movement.

The electrical connections of the servomotors pass through cable channels located in the upper shaft of the motor. The top of the duct is covered with a rubber seal to prevent water leakage. Compound joints shall be connected to each other by a series of chain structures. The flexible caudal fin is attached to the peduncle. The second linkage chassis is mounted on the lower and upper second link enclosures mounted on the first link. The electrical connections of the second

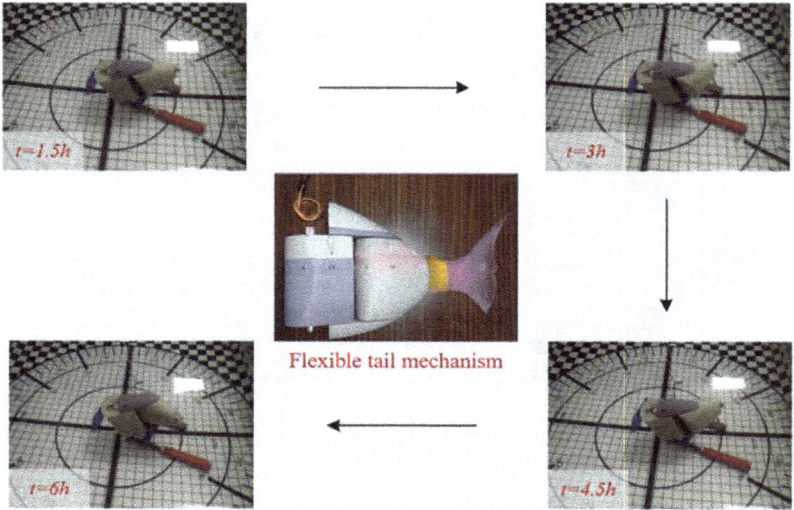

Figure 5.6. Test cycle of the tail mechanism for waterproofness performance during six hours.

linkage are also carried to the first linkage motor housing through the conduit in the second linkage housing. Here, the electrical connections of the servomotors are combined with a servo port circuit board and are transmitted to the front body electronic control system via the upper shaft of the first link. The upper and lower first link casings are located in the main body of the robot. As shown in Figure 5.6, this composite tail mechanism is analyzed during six hours in a swimming pool to test seal performance. Before and after experimental tests, the mass of the tail mechanism is measured with the sensitive scale. In both cases, it was observed that the measurements were equal.

The tail fin mold was also designed using 3D printing technology as shown in Figure 5.7. The white RTV-2 mold silicone is poured into the outer mold and the support is placed in the silicone.

The mechanical structure necessary to obtain the following sinusoidal joint angles is achieved.

$$\theta_1 = A_1 \sin(2\pi\, ft \pm \varphi_1), \tag{5.1}$$

$$\theta_2 = A_2 \sin(2\pi\, ft \pm \varphi_2). \tag{5.2}$$

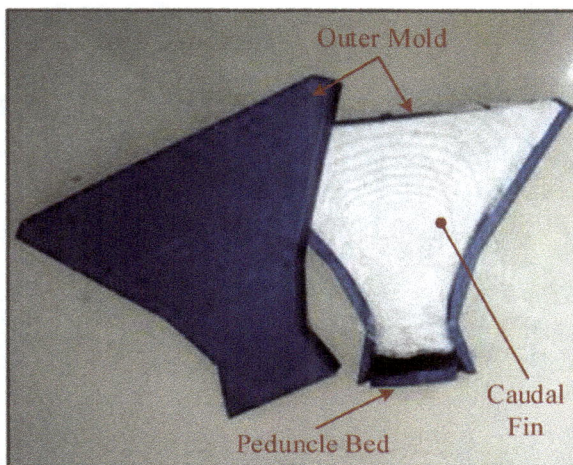

Figure 5.7. The produced caudal fin.

where A_1 and A_2 define the amplitude values of the first and second link angles; f indicates the flapping frequency of the tail mechanism; φ_1 and φ_2 describe the phase differences of the first and second link angles related to each other.

5.2.4. *CoG Control Mechanism Design*

Figure 5.8 shows the mechanical structure of a CoG control mechanism designed for 3D motion. The diving force generated by the CoG control mechanism in the main body enables the swimming robot to perform buoyancy and diving movements, one of the important swimming positions of the swimming robot, by rotating the robot in the y axis. The designed mechanism mainly consists of a single sliding block moving in the x axis with mechanical components and an AR3603HB RC servomotor. The x axis is in the same direction as the head of the swimming robot. The sliding block allows the change of CoG, that occurs as a prismatic lead table and a battery mounted on the table. The masses of the lead and battery table are about 300 g and 90 g, respectively. The RC servomotor is installed between the CoG motor housing and the main body. The T8 threaded spindle and nut system installed on the servomotor gear shaft is used to ensure

(a)

(b)

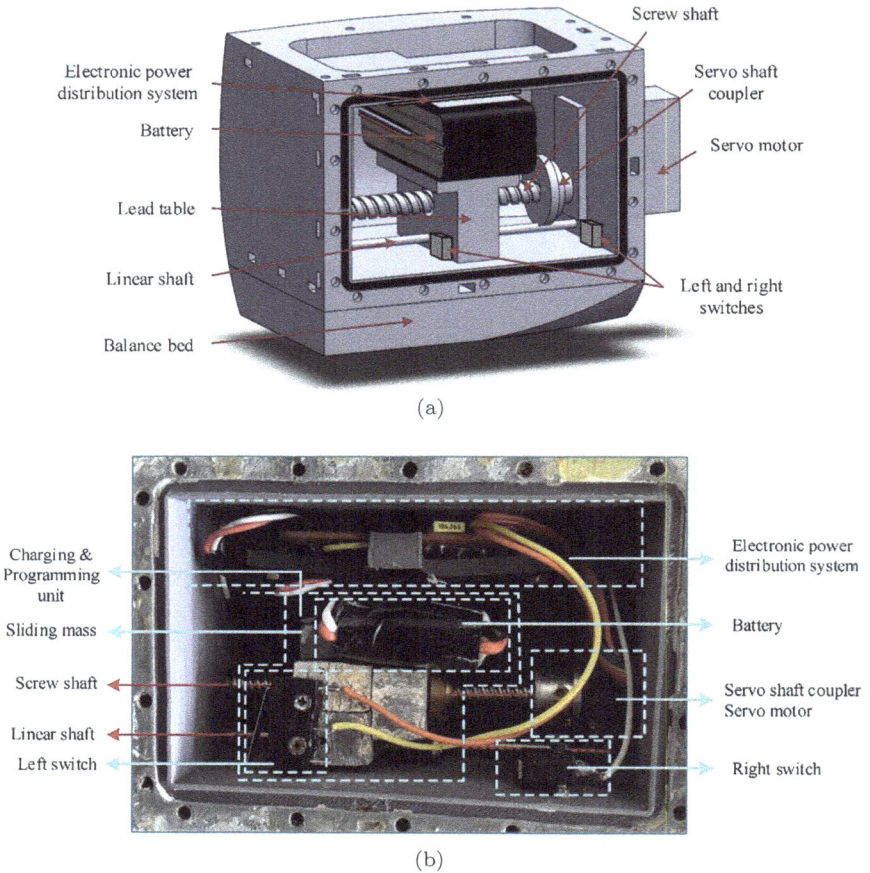

Figure 5.8. The developed CoG control mechanism: (a) Mechanical configuration. (b) Experimental parts.

the movement of the lead table. The rotation of the servomotor is converted into linear motion by the screw shaft. A fixed linear shaft is used in the same direction as the threaded spindle to prevent rolling effects that may occur during of table movement. Two boundary keys are placed on the left and right sides to define the bounds of the linear movement of the table along the x axis. The left switch is fixed to the table and the right switch is fixed to the main body in the opposite direction. An electronic power distribution system, to which the battery is connected and through which voltage levels are determined, is installed on the sliding block.

The control unit and the first common upper shaft are located above the main body. A cable channel is formed between the upper end and the main body for electronic connections. In order to ensure a seal between these two parts, the surfaces are covered with a fluid seal. This channel is fitted with a grease oil o-ring. The surfaces of other body parts are also fixed with a liquid sealing layer to increase water resistance.

The CoG position of the robot should be in the same direction as the center of the lift and slightly down. There is 120 g lead added in the balancing mass located between the main body and the first lower link housing. With this structure, balance is ensured in the vertical direction and the roll angle of the swimming robot becomes almost $0°$ while the swimming robot swims in the water.

In addition, the density of the swimming robot model is modeled at approximately $998\,\text{g/m}^3$, which is very close to the density of water. Thus, it is ensured that the swimming robot can easily swim in 3D space and upward to the surface of the water when the diving force is not applied. The following equation gives the expression necessary to determine the volume of the robotic fish Δf and the total mass m according to the desired mass,

$$\Delta f \cong \left.\frac{m}{\rho}\right|_{\rho \cong 998\,\text{g/m}^3} \qquad (5.3)$$

Finally, CoG control mechanism generated by the torque can be defined as,

$$\tau_x = m_x g(l_x \sin(\phi)\sin(\theta) - h_x \sin(\phi)\cos(\theta)). \qquad (5.4)$$

where g is the acceleration of gravity; l_x is the distance from the Center of Mass (CoM) of the prototype; h_x is the height from the CoM. θ and ϕ represent the pitch and roll angles, respectively.

5.2.5. *The Front View of Unit Design*

The front view module of the swimming robot model is designed with the sensors placed horizontally parallel to each other in the front

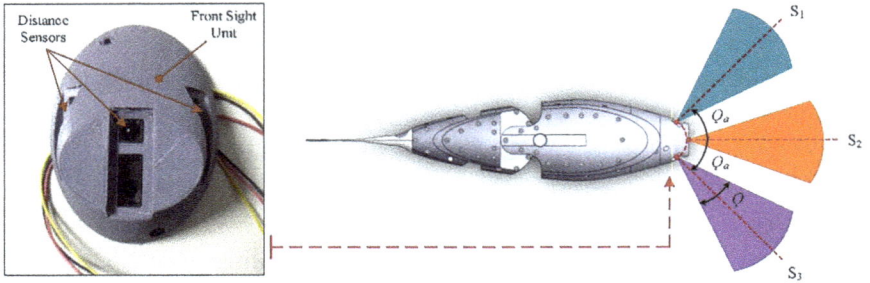

Figure 5.9. The designed front view unit and angles of sight.

axle so that the swimming robot can swim independently, avoiding obstacles in the water. As shown in Figure 5.9, this unit is divided into three sighting distances as left, right and frontal sections. The GP2Y0A21YK0F infrared distance sensor is used in each section.

The front sensor is positioned in the same direction as the x-axis and the left and right sensors are located with the front sensor at angular distances of $45°$ (Q_a). The linear measurement range of each sensor is from 6 to 80 cm, with S1, S2, and S3, which are the left, forward, and right distance sensors, respectively. The measuring range of these sensors is set to 6–60 cm to identify obstacles at acceptable time intervals in experimental studies. Here, Q is the obstacle detection angle at the sensor's viewing distance and is approximately $40°$ for robotic fish sensors.

5.2.6. *Electronic System Design of the Swimming Robot*

As shown in Figure 5.10, the electronic system of a swimming robot contains integrated software and hardware capable of energy, communication, cognition, and motion control functions. The electronic instrumentation system operates the mechanical parts and collects environmental data. According to the collected environmental data, the developed CPG control algorithm provides the required movement coordination of the swimming robot.

A 7.4 V 1350 mAh rechargeable Li-Po battery is used to power the robotic fish. Li-Po battery is smaller, lighter, lasts longer, and

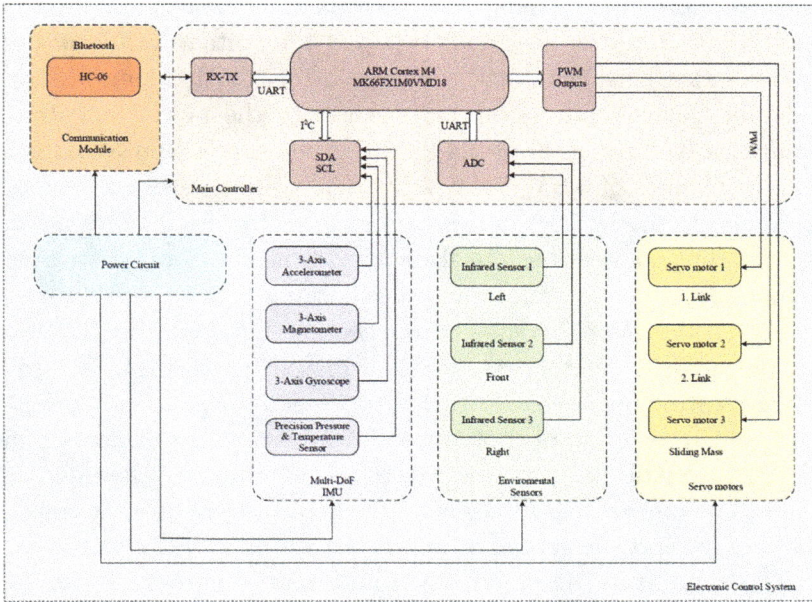

(a)

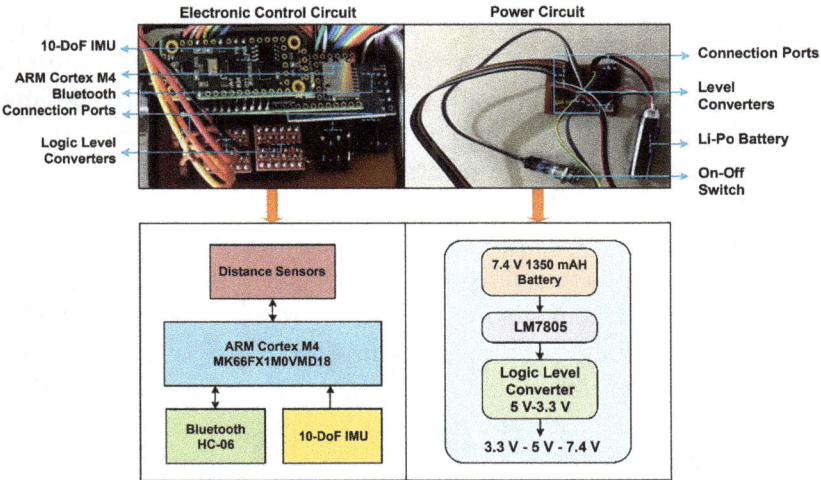

(b)

Figure 5.10. Electronic system of the robot: (a) Hierarchical block diagram of the system, which includes six parts: power supply, 10-DoF IMU, environmental sensors, servomotor system, microcontroller and communication module. (b) Printed circuits.

can store more energy than other rechargeable batteries. The power distribution circuit is designed to adjust the voltage values of other components where the supply voltage levels to the components of electronic devices differ from each other. In the power distribution circuit, the voltage level of 7.4 V of the battery is converted to the necessary values of 3.3 V and 5 V by logic voltage level converters.

In addition, the 16 mm waterproof on/off switch, charging, and programming connections are also moved from the power distribution circuit. Also, the programming and charging connection inputs are located in the main body. The connection between the Bluetooth module and the microcontroller is achieved by the RX/TX serial communication protocol. In order to detect obstacles on the left, right, and front of the robot, three infrared distance sensors Sharp GP2Y0A21YK0F are connected to the ADC inputs of the microcontroller. The mechanical movements of the prototype are provided by servomotors and the servomotors are driven by PWM signals.

The modularly designed electronic control system is an integrated circuit board comprising a microcontroller board, Prop Shield 10-DoF IMU, Bluetooth, and logic voltage level converters. This circle is located in the same direction as the x-axis of the robot. An IMU with a sampling frequency of 100 Hz is used to determine the roll, tilt and skew angles of the robot and to measure the linear position and velocity information in the three-dimensional space. At the same time, the environmental temperature can be measured. The IMU has a single 6-axis linear accelerometer FX0S8700CQ and a compass sensor, a 3-axis FXAS21002C gyroscope, and an MPL3115A2 barometric pressure and temperature sensor. Information is transferred from the IMU to the microcontroller using I2C, an asynchronous serial communication protocol, by (Serial Data) SDA and (Serial Clock) SCL. The microcontroller performs the rhythmic movements of the prototype using the intelligent control algorithm by evaluating the data received from the sensors and provides simultaneous continuity of these movements. There is a 32-bit 180 MHz ARM Cortex M4 MK66FX1M0VMD18 microprocessor on the control circuit. It has 1MB flash memory, 256KB RAM, and 4KB EEPROM internal memory.

The developed robotic fish prototype can swim independently and manipulate wireless information sent by the user's computer. In addition, near the surface of the water, sensor data can be simultaneously transmitted to the user's computer via a Bluetooth connection.

5.2.7. *Experimental Results and Discussion*

The experimental system has been set up to evaluate a swimming robot model as if in a real environment. In this system, the width of the pool is 200 cm, 300 cm long, and 66 cm deep, 1 Sony FDR-AX53 4K Demo Pool and GoPro Hero5 Black Edition Underwater Camera, 1 User PC with i7 2.4 GHz Processor, 8 GB DDR4 RAM, and 2 GB RAM Bytes. The graphics card is located as shown in Figure 5.11.

The diving motion response is shown in Figure 5.12. When the swimming robot reaches the steady-state, the junction angles become stable and a small amplitude is generated. For this experiment, first, the reference step angle is kept at 15° for 4 s, and then at the end of the 4 s period, this angle is assumed to be at 0° to stay level. The snapshots obtained from the underwater action camera are located in the middle of the long edge of the pool.

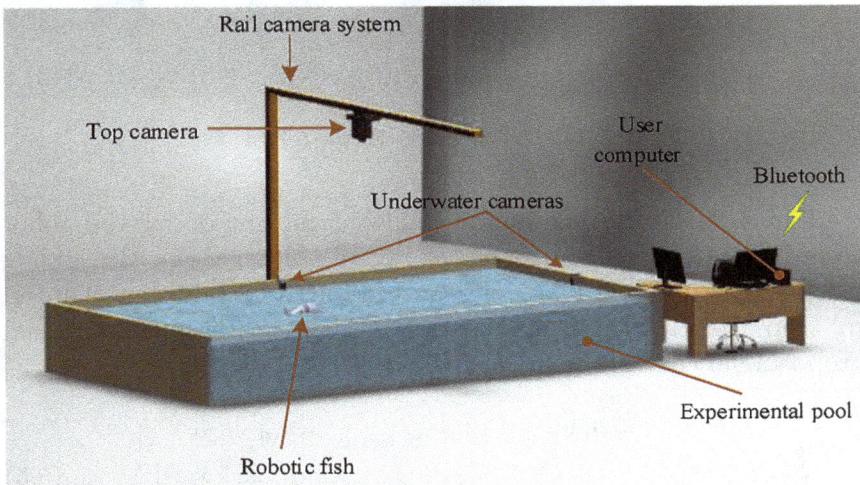

Figure 5.11. Experimental setup.

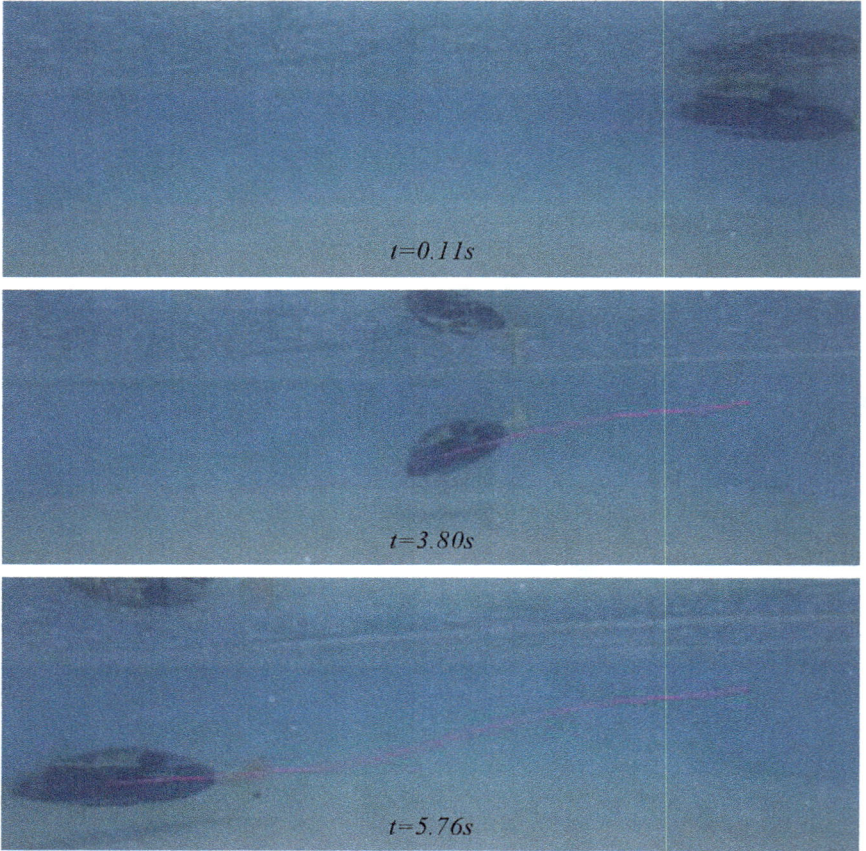

Figure 5.12. Diving motion of the robotic fish.

Figure 5.13 also shows the angle of inclination of the robot according to the reference step angle. In this analysis, the sliding block is moved forward and backward in the CoG control mechanism to keep the tilt angle at the desired level with a conventional proportional control structure as follows,

$$\left\{ \begin{array}{c} \text{if } \theta_{\text{ref}}(k) - \theta(k) > 0 \text{ sliding}_{\text{mass}} \text{ goes forward} \\ \text{else if } \theta_{\text{ref}}(k) - \theta(k) > 0 \text{ sliding}_{\text{mass}} \text{ goes backward} \\ \text{else sliding}_{\text{mass}} \text{ keeps position} \end{array} \right\} . \quad (5.5)$$

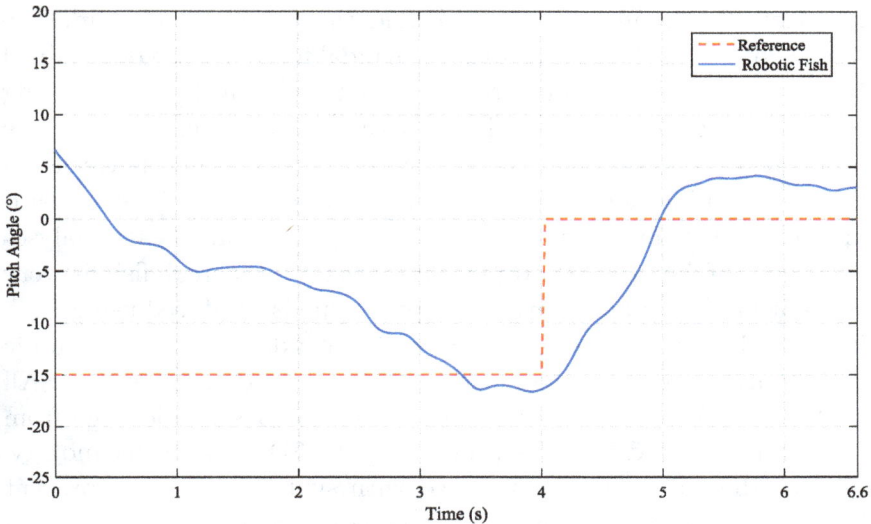

Figure 5.13. The pitch angle of the robotic fish prototype according to the reference.

It is evident from Figure 5.13 that the control mechanism developed in CoG is efficient enough to maintain the prototype at the desired pitch angle.

5.2.8. *Conclusions*

This study presents the biomimetic design and prototype fabrication of an intelligent robotic fish (i-RoF) based on life-inspired swimming to perform real-world exploration and survey tasks. The developed swimming robot mimics the swimming poses of the Carangiform BCF type with a two-link tail mechanism. In addition, a CoG control mechanism for 3D swimming performance was developed. The CPG-based motion controller was adapted to generate powerful, smooth, and rhythmic oscillating swimming patterns. The proposed CPG model is designed to serve as the lamprey spinal cord to ensure intelligent control. The swimming robot model was examined for 3D locomotion capabilities in the real experimental system.

The swimming robot consists of five basic components including the main body, the control unit, the front view unit, the two-link propulsion mechanism of the tail, and the flexible caudal fin. The dimensions of the flexible caudal and caudal fin bonds are determined by analyzing 50 swimming patterns of the forward and cornering movements of true carps. The obtained correlation lengths are adapted to the designed model. Robotic swimming patterns of fish in a SimMechanics environment are also analyzed for forward and cornering movements. It can be seen from these analyses that quantifying the prototype according to the real fish is very convenient. Every hard part of the prototype is produced using PLA in 3D printing technology. The flexible caudal fin of the robotic fish model is also produced using a white-colored silicone mold. All parts are covered with epoxy resin to prevent possible leakage from micropores formed in parts produced with 3D printing technology. Finally, the outer surface is coated with synthetic paint to prevent leaks that may result from capillary cracks during assembly. In order to test the sealing performance of the composite parts, the model is worked for 6 hours in a test basin filled with water. The success of the sealing tests was noted.

To validate the diving motion, two-pitch reference angles are created and a sufficiently effective depth performance is obtained. According to experimental studies, the swimming displays of the developed prototype are very effective for real-world exploration and survey tasks with their new fish-like design.

5.3. Diving System of Labriform Swimming Robot

In this section, Labriform swimming robot as glider will be introduced, in which, the swimming robot has two types of independent actuators: the first one is the sliding shaft of the center of gravity (CoG), and the second one is used for pectoral fin actuation. The full-body dynamics of the robot diving system are derived, in which a computational fluid dynamics (CFD) from SOLIDWORKS® is adopted to show the relationship of drag, lift, and moment with the angle of attack variation. Several practical experiments have been conducted in order to investigate the success of the proposed model and compared with simulated results. The results showed a great match between the practical and simulation results and the success of the suggested model.

5.3.1. *Diving System of Swimming Robot*

The diving module presented here consists of one stepper motor and a movable mass sliding on the lead screw of the motor. Motion commands are sent to the microcontroller unit via the HC-05 Bluetooth module. In order to overcome the drag forces generated due to diving and the use of the largest surface area fins, a set of four 1.5 V AA batteries were used to supply energy to yaw and pitch motors, communication unit and the microcontroller of the robot with the required energy as shown in Figure 5.14. The robot's body is set to be first at neutrally buoyant state. Then after a specified period the pitch motor starts to rotate allowing the movable mass to be shifted toward the frontal part of the body. This will in turn pitch the robot down within an angle with the horizontal plane of the body. Since the yaw motor is triggered to operate synchronously, the robot will start to dive gradually. After a pre-determined period, the pitch motor will start to roll back, and consequently the movable mass will be shifted back towards the center of gravity (CoG) point to allow the robot to float gradually and then reach the neutrally buoyant state again. Further details are discussed in the next sections.

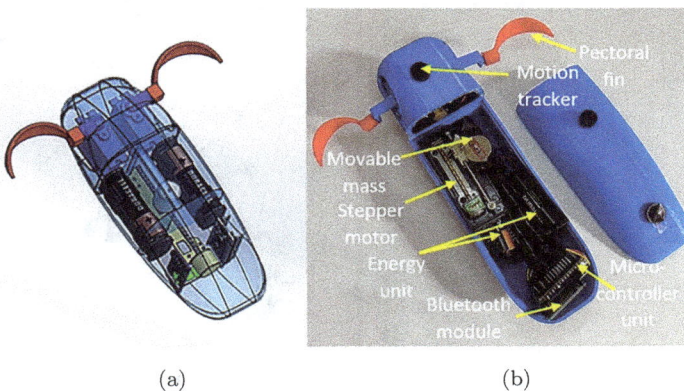

(a)   (b)

Figure 5.14. Swimming robot design. (a) SOLIDWORKS® model. (b) Physical model.

5.3.2. *Modeling of Diving System*

The mass sliding approach is designed with a combinations of linear motion motor, 5 V stepper motor defined as (pitch motor), and a sliding mass with a specified weight fixed onto linear motion shaft as shown in Figure 5.15. During the sliding movement of the movable mass forwards or backward to the stepper lead screw, the proposed mass-gravity model of the whole body also goes forwards or backward, respectively. This way the robot pitches down or up; under the fins propulsion, the robot's body will be pushed downward or upward, respectively. The forces act on the sliding mass as follows:

The weight of the moving mass $P = M_m g$ (M_m and g are the movable mass and gravitational acceleration, respectively).

(1) The normal force $F_{\text{norm}} = P_2$ (P_2 is the vertical components of P; $P_2 = P \cos \theta$) and $\theta°$ is the pitch angle with the horizontal axis)
(2) The total frictional force $F_{\text{fric}} = \alpha P_2 + f$ (where α and f are frictional coefficient and resistance of guide surface without load, respectively).
(3) The force of the pitch motor $F_m = P_1 = -P \sin \theta^*$.

With an assumption that the movable mass is sliding (forward/backward) within a speed of $\dot{X}_m$, and when the robot pitches

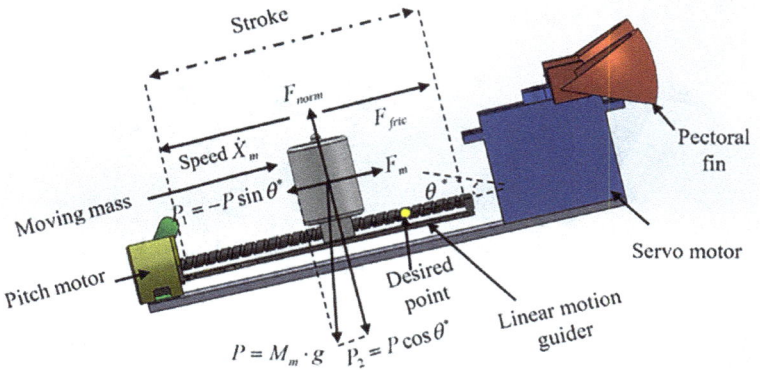

Figure 5.15. Mass shifter mechanism.

(upward/downward) it will make an angle θ^* with the horizontal plane. If the mass slides toward the rear part of the body, the friction is always opposite to the speed direction, then the total friction direction is toward the frontal part of the robot. The dynamics of the movable mass can be described by the following differential equation:

$$M_m \cdot \ddot{X}_m = F_{\text{fric}} + F_m \tag{5.6}$$

where $\ddot{X}_m$ can be defined as the acceleration of the movable mass. The body reference frame is fixed with the origin of the body at its center of buoyancy. This reference frame is described relatively to the global fixed earth reference frame as shown in Figure 5.16. It is assumed that the robot is initially neutrally buoyant, ignoring surface effects on the robot and underwater currents. The position of the robot and its orientation relative to the earth reference frame are given as follows: (x^*, y^*, z^*) and $(\theta^*, \phi^*, \psi^*)$, respectively.

5.3.2.1. *Swimming robot kinematic model*

The position of the robot is given by the vector $[x^*, y^*, z^*]^T$ and the orientation is defined by the vector $[\phi^*, \theta^*, \psi^*]^T$ and can be

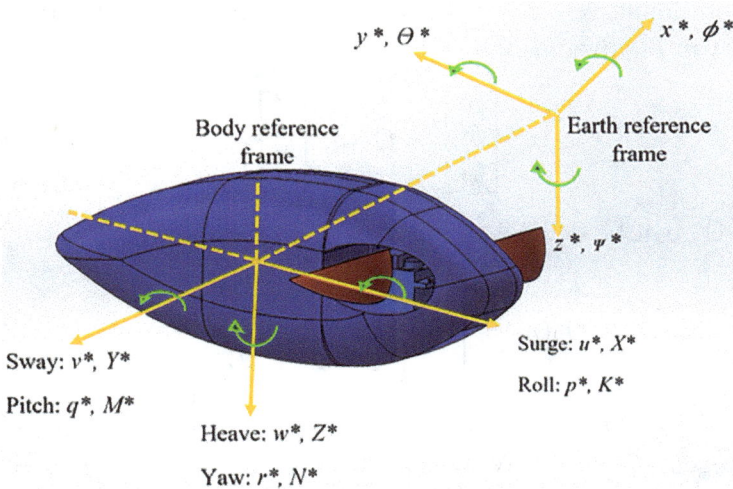

Figure 5.16. Body and earth global coordinate systems.

described relative to the global earth coordinate. The translational motions between the body and earth global coordinates can be given as follows:

$$
\begin{bmatrix} \dot{x}^* \\ \dot{y}^* \\ \dot{z}^* \end{bmatrix} = \mathbf{O}_1(\eta) \begin{bmatrix} u^* \\ v^* \\ w^* \end{bmatrix}
\tag{5.7}
$$

where $\eta = (\varphi^*, \theta^*, \psi^*)$ and $\mathbf{O}_1$ is defined as follows:

$$
\mathbf{O}_1(\eta) = [\mathbf{O}_1^1(\eta)\ \mathbf{O}_1^2(\eta)\ \mathbf{O}_1^3(\eta)]
\tag{5.8}
$$

$$
\mathbf{O}_1^1(\eta) = \begin{bmatrix} c\psi^* c\theta^* \\ s\psi^* c\theta^* \\ -s\theta^* \end{bmatrix}
\tag{5.9}
$$

$$
\mathbf{O}_1^2(\eta) = \begin{bmatrix} -s\psi^* c\phi^* + c\psi^* s\theta^* s\phi^* \\ c\psi^* c\phi^* + s\psi^* s\theta^* s\phi^* \\ c\theta^* s\phi^* \end{bmatrix}
\tag{5.10}
$$

$$
\mathbf{O}_1^3(\eta) = \begin{bmatrix} s\psi^* s\phi^* + c\psi^* s\theta^* c\phi^* \\ -c\psi^* s\phi^* + s\psi^* s\theta^* c\phi^* \\ c\theta^* c\phi^* \end{bmatrix}
\tag{5.11}
$$

The rotational velocities between the body and earth reference frames can be given as follows:

$$
\begin{bmatrix} \dot{\phi}^* \\ \dot{\theta}^* \\ \dot{\psi}^* \end{bmatrix} = \mathbf{O}_2(\eta) \begin{bmatrix} p^* \\ q^* \\ r^* \end{bmatrix}
\tag{5.12}
$$

where $\mathbf{O}_2$ can be defined as follows:

$$
\mathbf{O}_2(\eta) = \begin{bmatrix} 1 & s\phi^* t\theta^* & c\phi^* t\theta^* \\ 0 & c\phi^* & -s\phi^* \\ 0 & s\phi^*/c\theta^* & c\phi^*/c\theta^* \end{bmatrix}
\tag{5.13}
$$

where s, c and t refer to the trigonometric functions of sine, cosine and tangent, respectively. It is good to mention that the robot will not experience singularity at $\theta^* = \pm 90°$, since the pitch motion will be limited to $\pm 90°$ for better performance.

5.3.2.2. *Swimming robot dynamic model*

Assuming the glider-robot's body reference frame is centered at its buoyancy center, when the moving mass is shifted, it produces very small change in the moments of inertia I_{xx}^*, I_{yy}^*, and I_{zz}^*, and are constants. The product of inertia I_{xy}^*, I_{xz}^*, and I_{yz}^* are small in comparison to the moments of inertia and can be neglected as follows:

$$I^* = \begin{bmatrix} I_{xx}^* & 0 & 0 \\ 0 & I_{yy}^* & 0 \\ 0 & 0 & I_{zz}^* \end{bmatrix} \tag{5.14}$$

Then motion equations in 6-DOF with respect to the body reference frame can be described as in:

$$\left. \begin{aligned} & m^*[\dot{u}^* - v^*r^* + w^*q^* - x_g(q^{*2} + r^{*2}) + y_g(p^*q^* - \dot{r}^*) \\ & \quad + z_g(p^*r^* + \dot{q}^*)] = \sum X^* \\ & m^*[\dot{v}^* - w^*p^* + u^*r^* - y_g(r^{*2} + p^{*2}) + z_g(q^*r^* - \dot{p}^*) \\ & \quad + x_g(q^*p^* + \dot{r}^*)] = \sum Y^* \\ & m^*[\dot{w}^* - u^*q^* + v^*p^* - z_g(p^{*2} + q^{*2}) + x_g(r^*p^* - \dot{q}^*) \\ & \quad + y_g(r^*q^* + \dot{p}^*)] = \sum Z^* \\ & I_{xx}^*\dot{p}^* + (I_{zz}^* - I_{yy}^*)q^*r^* + m^*[y_g(\dot{w}^* - u^*q^* + v^*p^*) \\ & \quad - z_g(\dot{v}^* - w^*p^* + u^*r^*)] = \sum K^* \\ & I_{yy}^*\dot{q}^* + (I_{xx}^* - I_{zz}^*)r^*p^* + m^*[z_g(\dot{u}^* - v^*r^* + w^*q^*) \\ & \quad - x_g(\dot{w}^* - u^*q^* + v^*p^*)] = \sum M^* \\ & I_{zz}^*\dot{r}^* + (I_{yy}^* - I_{xx}^*)p^*q^* + m^*[x_g(\dot{v}^* - w^*p^* + u^*r^*) \\ & \quad - y_g(\dot{u}^* - v^*r^* + w^*q^*)] = \sum N^* \end{aligned} \right\} \tag{5.15}$$

where

- u^*, v^*, and w^* are the linear velocity components known as surge, sway and heave, respectively.
- p^*, q^*, and r^* are the angular velocity components known as roll, pitch and yaw, respectively.

- X^*, Y^*, and Z^* are the external forces
- K^*, M^*, and N^* are external moments.

The center of gravity of the robot is defined as c_G and the center of buoyancy is defined as c_b and they can be given as the following:

$$c_G = [x_g y_g z_g]^T \tag{5.16}$$

$$c_b = [x_b^* y_b^* z_b^*]^T \tag{5.17}$$

where, x_g, y_g, z_g are the center of gravity relative to the origin at center of buoyancy, x_b^*, y_b^*, z_b^* are the center of buoyancy at the origin of body's reference frame, and m° is the total mass of the robot.

The robot experiences a hydrostatic force/moment due to the weight of the robot and buoyancy force. However, the weight can be defined as $W = m^* g$, while the buoyancy is given as $B = \rho \vee g$, where g is the gravitational force and $\vee$ is the total volume of the displaced water.

The hydrostatic forces/moments can be given as:

$$\left. \begin{aligned} \mathbf{F}_S &= \mathbf{f}_G - \mathbf{f}_B \\ \mathbf{M}_S &= \mathbf{c}_G \times \mathbf{f}_G - \mathbf{c}_B \times \mathbf{f}_B \end{aligned} \right\} \tag{5.18}$$

where $\mathbf{F}_S$ and $\mathbf{M}_S$ are the hydrostatic force and moments, respectively $\mathbf{f}_G$ and $\mathbf{f}_B$ are the hydrostatic matrices due to the gravity and buoyancy, respectively. Expanding on these equations will result in the following nonlinear terms:

$$\left. \begin{aligned} X_S &= -(W - B)\sin\theta^* \\ Y_S &= (W - B)\cos\theta^* \sin\phi^* \\ Z_S &= (W - B)\cos\theta^* \cos\phi^* \\ K_S &= -(y_g W - y_b B)\cos\theta^* \cos\phi^* - (z_g W - z_b B)\cos\theta^* \sin\phi^* \\ M_S &= -(z_g W - z_b B)\sin\theta^* - (x_g W - x_b B)\cos\theta^* \cos\phi^* \\ N_S &= -(x_g W - x_b B)\cos\theta^* \sin\phi^* - (y_g W - y_b B)\sin\theta^* \end{aligned} \right\}$$

$$\tag{5.19}$$

The proposed design of the glider-robot uses two control systems for both vertical and horizontal motions. The pitch motion is controlled in an independent manner from the yaw motion. In order to design a diving system, only the depth plane will be taken into consideration. With the assumption that only x_g is considered, y_g and z_g are very small and can be neglected.

The robot's kinematics between the earth and body reference frames can be simplified as follows:

$$\left.\begin{aligned} \dot{x}^* &= u^* \cos\theta^* + w^* \sin\theta^* \\ \dot{z}^* &= -u^* \sin\theta^* + w^* \cos\theta^* \\ \dot{\theta}^* &= q^* \end{aligned}\right\} \qquad (5.20)$$

Linear velocity component u° is constant and the heave component w° is small and can be ignored. Thus the robot kinematics will be further simplified to:

$$\left.\begin{aligned} \dot{x}^* &= u^* \cos\theta^* \\ \dot{z}^* &= -u^* \sin\theta^* \\ \dot{\theta}^* &= q^* \end{aligned}\right\} \qquad (5.21)$$

Furthermore, the robot's rigid body dynamics given in equation (5.15) can be simplified to only the depth plane. All terms (v^*, p^*, r^*) that are out of the depth plane will be set to zero, dropping any higher order terms.

The center of gravity due to the x^*-component x_g can be given as the following:

$$x_g = \frac{M_m \cdot X_m}{m^*} \qquad (5.22)$$

where M_m is defined as the movable sliding mass and X_m is the sliding x^*-component coordinate of the moving mass.

The pitch motor is chosen to be a "stepper motor" as it is very useful and performs a significant role in the control process with precise control of speed, angular position and direction with open loop systems.

In order to obtain inertia coefficients and other parameters concerning the robot's body, the same equations given in Chapter 3 (equations (3.15)–(3.26)) were used.

5.3.3. *Simulation and Experimental Results*

5.3.3.1. *Simulation results*

The robots aim to swim at different angles of attacks as shown in Figure 5.17. When the robot swims, the water velocity around the body will be distributed as shown in Figure 5.18, the frontal part of the robots will act as the stagnation point of the robot as the velocity of the surrounding water is low and hence the static pressure will have a maximum magnitude. Moreover, it is noted that the pectoral fins have high pressure acting on them than the rest of the robot's body as shown in Figure 5.19.

This difference in pressure between the front and the back of fins gives rise to the drag force. The relevant drag force results obtained from SOLIDWORKS® fluid flow simulations are displayed in Table 5.2. The negative sign (−) before each angle of attack refers to the direction of swimming downward (i.e. during diving), whereas the unsigned angles refer to the direction upward (i.e. during floating).

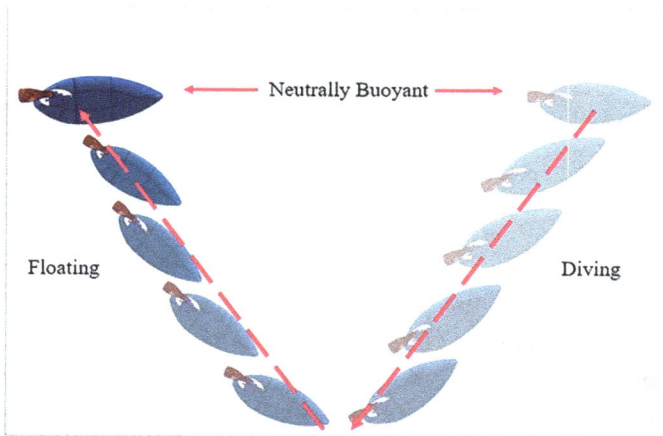

Figure 5.17. Path of the swimming robot.

(a)

(b)

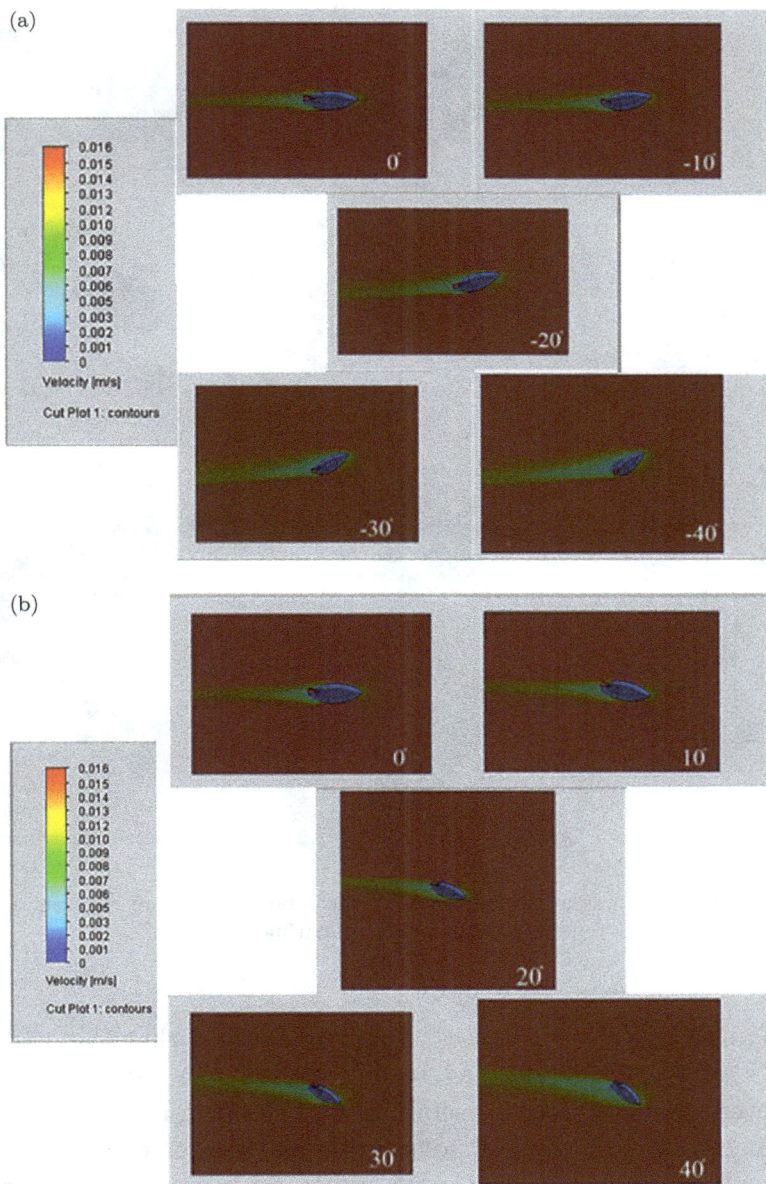

Figure 5.18. The velocity of water around the robot at different angle of attacks (isometric view). (a) Diving, (b) Floating.

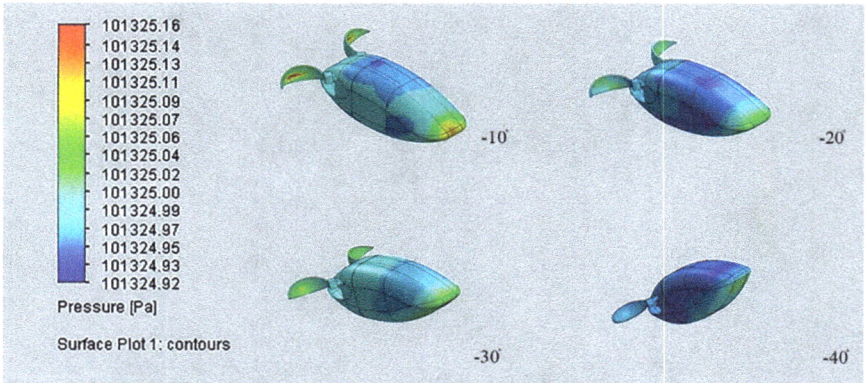

(a)

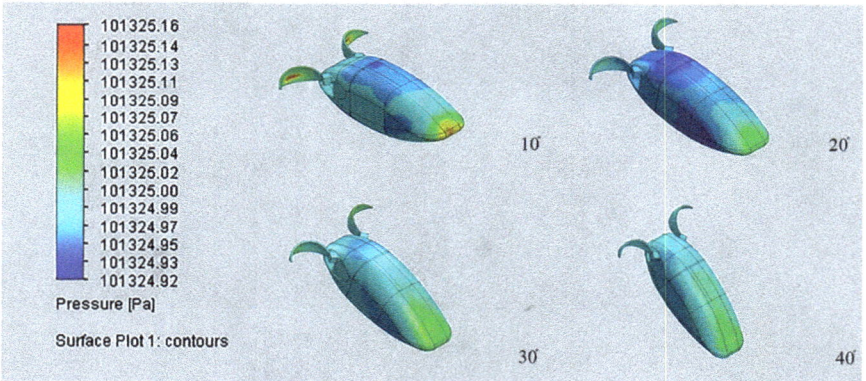

(b)

Figure 5.19. The pressure distribution on the robot's surface at different angles of attacks (isometric view). (a) Diving. (b) Floating.

From the previous figures, it can be noticed that the angle of attack of the body will lead to an increasing in the drag forces, however, this drag force is useful for the swimming robot. Increasing the angle of attack will lead to a significant drag force. For instance, it would help in stabilizing the robot when it is subjected to large moments. The drag coefficient at different angles of attacks is calculated as shown in Figure 5.20. It can be noticed that the great effect of the variation of angle of attack on the drag coefficient,

Table 5.2. Drag force simulation results (N).

Angle of attack	Final value
0°	2.35E-03
−10°	5.36E-04
−20°	1.50E-03
−30°	3.72E-03
−40°	5.47E-03
10°	2.54E-03
20°	3.11E-03
30°	4.20E-03
40°	5.19E-03

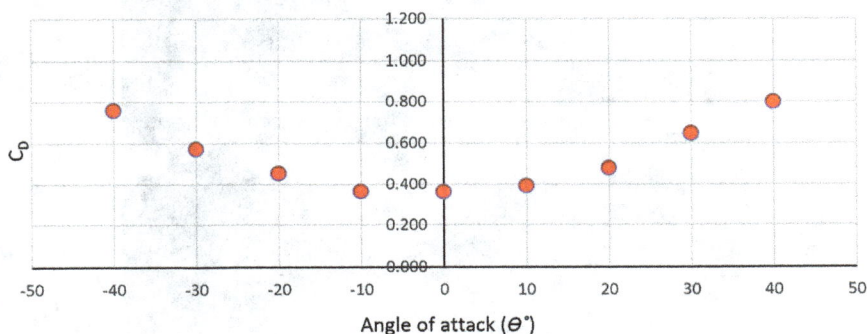

Figure 5.20. Drag coefficient at different angle of attacks.

such that increasing the angle of attack from 0° to 40° will lead to a significant effect on the drag coefficient and then the generated drag forces.

5.3.3.2. *Practical experimental results*

A swimming environment is designed to test the robot's ability to dive represented by a glass pool with dimensions of 100 cm, 30 cm and 45 cm of length, width and height, respectively. The transparent type of glass was chosen to capture the side motion of the robot during diving. A side camera was placed at a distance of 0.5 m from the swimming pool to record the movement of the robot to be analyzed

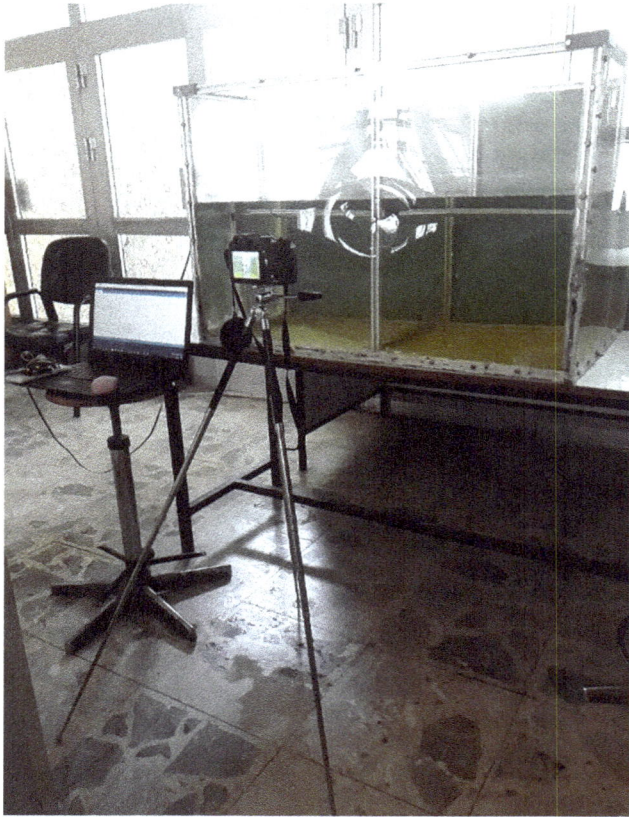

Figure 5.21. Swimming robot environment.

with an image processing toolbox in the Matlab platform, as shown
in Figure 5.21. The concave fin (Fin1) that was explained previously
in Chapter 2 was used during this experiment as it produces the
highest thrust.

The Air cavity inside the housing body helps to approximate the
density of the robot to the density of water by reducing the weight of
the swimming robot while maintaining its same initial volume. The
internal area also provides a secure, waterproof space for the control
system electronics and the propulsion mechanism. The unexpected
change of position of the components can lead to a change of the
location of the center of gravity and therefore to major stability

problems. Due to the small size of the suggested gliding robot, the power unit required to overcome the drag force has been added just below the center of gravity and fixed at the external surface of the body. An extra 50 gm weight has been added near the rear part of the robot to increase the mass of the robotic fish to achieve neutral buoyancy state. To investigate the mass shifter mechanism as a glider robot, a movable mass of 100 gm weight has been chosen to slide over the lead screw of 5.2 cm in length attached to the stepper motor from ROHS — model SM15L. The pitch motor is programmed to follow a sawtooth pattern as described previously. The total weight force of the robot W is 9.62 N and the buoyancy B is set to be a little bit larger than W of 9.94 N to ensure that the glider robot will float in case of failure.

For diving, it consumes 1 sec to move the movable mass from x_g position to reach the end of the lead screw. The lead screw step angle is 18° resulting in 20 steps/revolution with a pitch distance of 2 mm per revolution, so in order to reach the other end of the lead screw; it needs 26 revolutions and 520 steps. As the pitch motor starts to rotate from x_g toward the other end of the lead screw, the movable mass will consequently pitch the robot down and produce an angle θ^*. As the pitch motor continues to rotate, θ^* will increase linearly. Due to the electronics specifications (pitch and yaw motors) being unable to withstand higher pressure underwater, the pitch motor is programmed to achieve 40° of pitch angle for both diving and floating states as shown in Figures 5.22 and 5.23, respectively.

When the movable mass reaches the other end of the lead screw, the yaw motor is triggered and operates for 10 sec at highest power to stroke ratio of $R_F = 3:1$ as shown in Figure 5.24.

After 10 sec, when the robot reaches about 10 cm from the bottom of the pool, it is delayed for 1 sec to allow the pitch motor to rotate back to its x_g position. Then it continues to work for another 10 sec as shown in Figure 5.25 to achieve floating state. Both diving and floating frames are given in Figures 5.26(a) and 5.26(b), respectively.

Figure 5.27 shows a comparison between simulation and experimental results. The glider robot was able to achieve about 40° of pitch

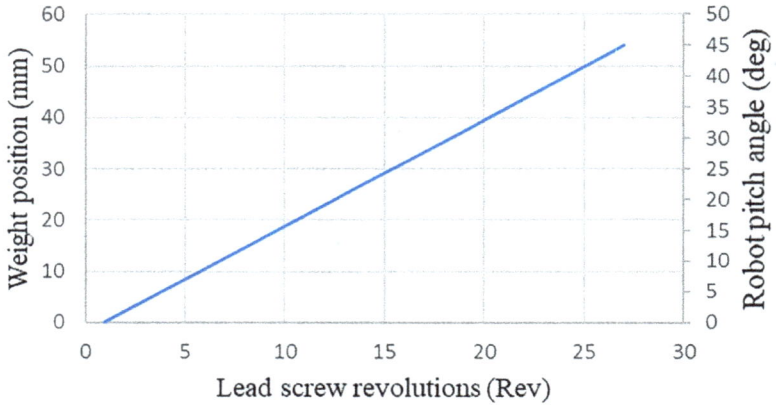

Figure 5.22. Pitch motor during diving.

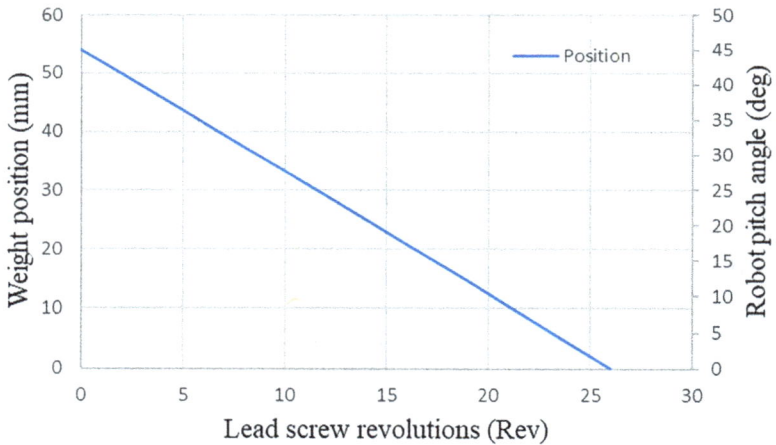

Figure 5.23. Pitch motor during floating.

angle and swim for a depth of approximately 30 cm underwater as shown in Figure 5.28.

5.3.4. *Conclusion*

In this section, the comprehensive development and application of a nonlinear mathematical model of the swimming robot as a glider are produced, the diving motion of mimicking Labriform swimming

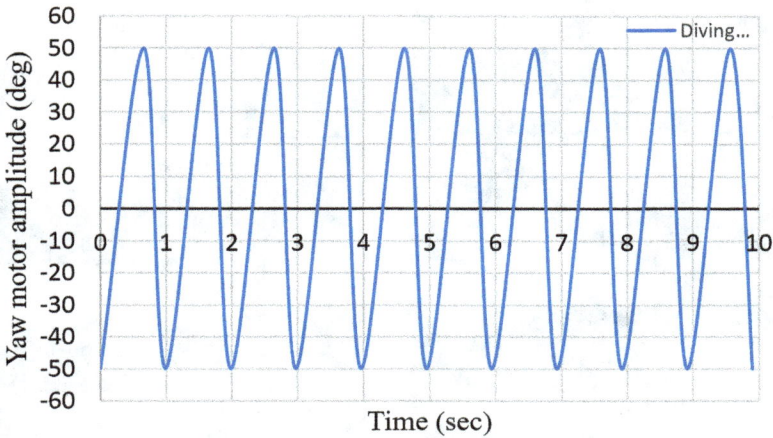

Figure 5.24. Yaw motor during diving.

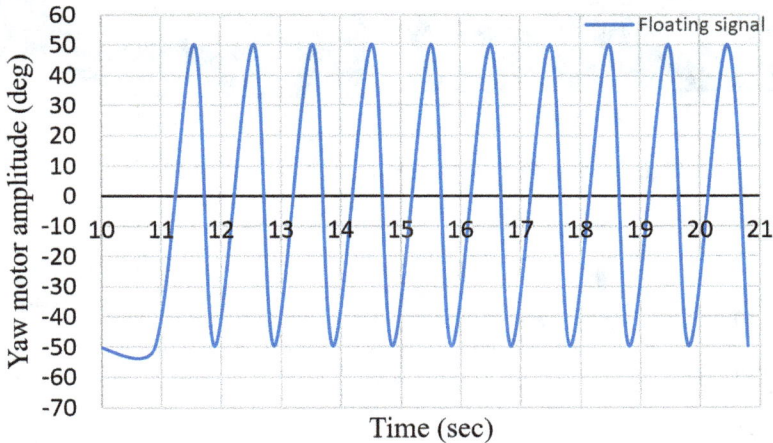

Figure 5.25. Yaw motor during floating.

fish is analyzed. The suggested robot used a mass sliding mechanism and reported to our best knowledge the smallest swimming-gliding robot that has been provided in the literature. The proposed glider-swimming robot has two types of independent actuators: The first one to realize the shaft of the center of gravity, and the second one used for pectoral fin actuation. The glider dynamic model presented

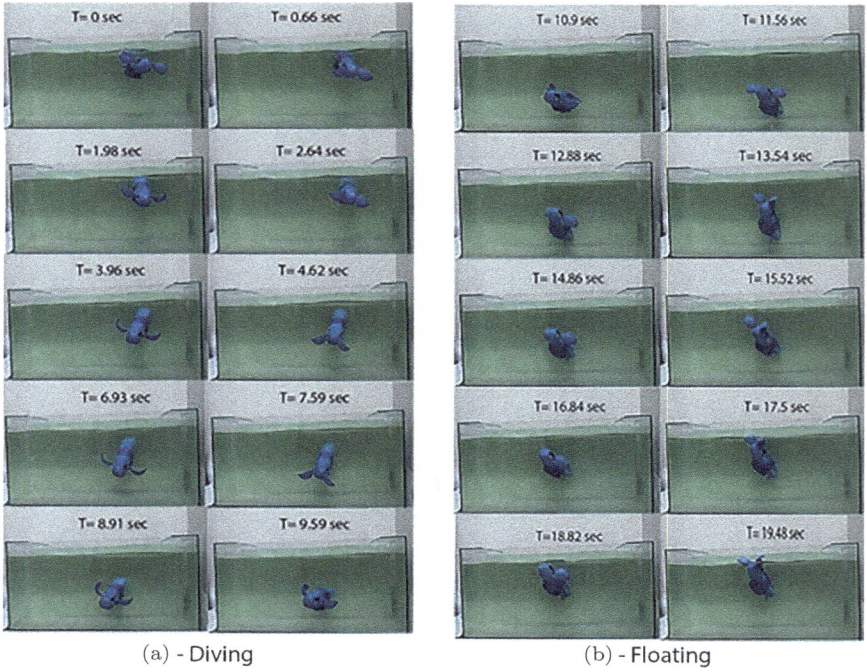

(a) - **Diving**

(b) - **Floating**

Figure 5.26. Swimming robot. (a) Diving. (b) Floating.

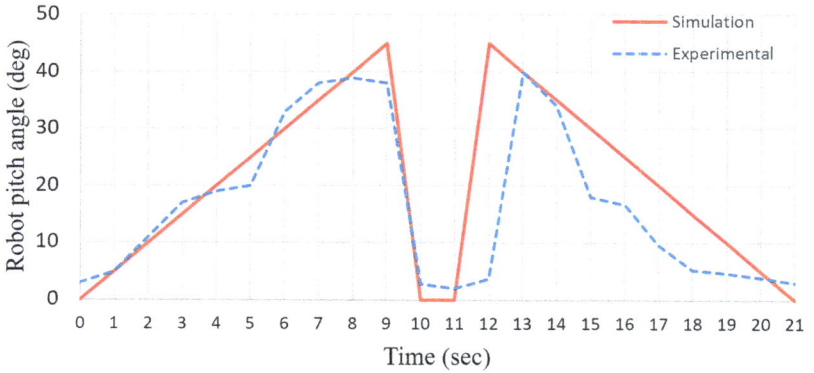

Figure 5.27. One cycle of diving and floating of the glider robot.

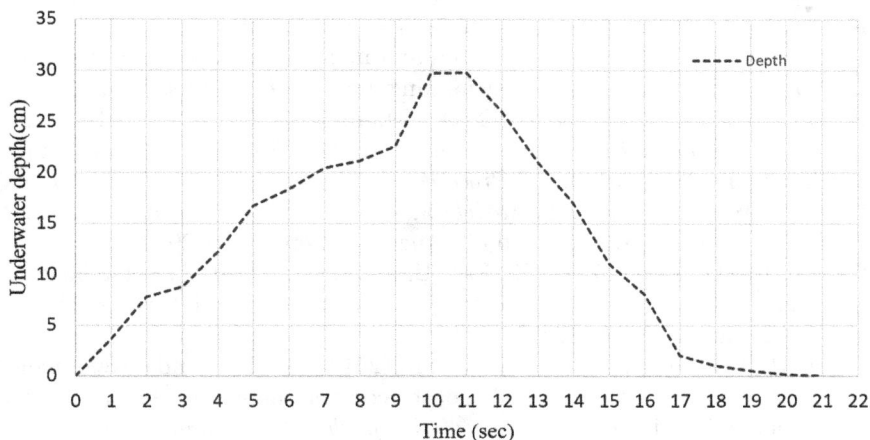

Figure 5.28. The achieved depth underwater.

here is general and not vehicle specific, and has applications in glider design, control, estimation, and optimization. It applies to underwater gliders with fixed mass and movable mass M_m which can control their buoyancy and center of gravity.

The robot's body is set to be first at neutrally buoyant state. Then after a specified period, the pitch motor starts to rotate allowing the movable mass to be shifted toward the frontal part of the body. This will in turn pitch the robot down within an angle. Since the yaw motor is triggered to operate synchronously, the robot will start to dive gradually. After a pre-determined period, the pitch motor starts to roll back, and consequently, the movable mass will be shifted back towards the C_G point to allow the robot to float gradually and then reach the neutrally buoyant state again. The results show the ability of the robot to dive and float following a predetermined sawtooth pattern.

References

Arima, M., Tonai, H. and Kosuga, Y. (2013). Underwater glider "SOARER" for ocean environmental monitoring, *In: Proceedings of the International Underwater Technology Symposium (UT)*, Tokyo, Japan, 5–8 March 2013, pp. 1–5.

Ay, M., Korkmaz, D., Ozmen Koca, G., Bal, C., Akpolat, Z. H. and Bingol, M. C. (2018). Mechatronic design and manufacturing of the intelligent robotic fish for bio-inspired swimming modes, *Electronics*, 7(7), pp. 118.

Bal, C., Korkmaz, D., Koca, G. O., Ay, M. and Akpolat, Z. H. (2016). Link length optimization of a biomimetic robotic fish based on big bang-big crunch algorithm, *In Proceedings of the 2016 21st International Conference Methods and Models in Automation and Robotics, MMAR 2016*, Miedzyzdroje, Poland, 29 August–1 September 2016.

Cao, J., Lu, D., Li, D., Zeng, Z., Yao, B. and Lian, L. (2019). Smartfloat: A multimodal underwater vehicle combining float and glider capabilities, *IEEE Access*, 7, pp. 77825–77838.

Claus, B., Bachmayer, R. and Cooney, L. (2012). Analysis and development of a buoyancy-pitch based depth control algorithm for a hybrid underwater glider, *In: Proceedings of the Autonomous Underwater Vehicles (AUV)*, Southampton, pp. 1–6.

Drężek, K. (2019). CFD approach to modelling hydrodynamic characteristics of underwater glider, *Transactions on Aerospace Research*, 4(257), pp. 32–45.

Eriksen, C. C., Osse, T. J. and Light, R. D. (2001). Seaglider: A longrange autonomous underwater vehicle for oceanographic research, *IEEE J. Oceanic Eng.*, 26, pp. 424–436.

Fossen, T. I. (2011). *Handbook of Marine Craft Hydrodynamics and Motion Control*, 1st Ed., Norway, John Wiley & Sons Ltd., pp. 122.

Graver, J. G. and Leonard, N. E. (2001). Underwater glider dynamics and control, *In: Proceedings of the 12th International Symposium on Unmanned Untethered Submersible Technology*, Durham, NH, pp. 1710–1742. https://doi.org/10.3390/electronics7070118

Isa K. K. and Arshad, M. R. (2011). Dynamic modeling and characteristics estimation for USM underwater glider, *In: Proceedings of the Control and System Graduate Research Colloquium (ICSGRC)*, Shah Alam, Malaysia, pp. 12–17.

Ji, D., Choi, H., Kang, J., Cho, H., Joo, M. and Lee, J. (2019). Design and control of hybrid underwater glider, *Advances in Mechanical Engineering*, 11(5), pp. 1–9.

Kato, N. (2000). Control performance in the horizontal plane of a fish robot with mechanical pectoral fins, *IEEE Journal of Oceanic Engineering*, 25(1), pp. 121–129.

Koca, G. O., Bal, C., Korkmaz, D., Bingol, M. C., Ay, M., Akpolat, Z. H. and Yetkin, S. (2008). Three-dimensional modeling of a robotic fish based on real carp locomotion, *Appl. Sci. 2018*, 8, p. 180.

Koca, G. O., Yetkin, S., Ay, M., Bal, C. and Akpolat, Z. H. (2017). FSI analysis of carangiform three dimensional multi-link biomimetic robotic fish, *AKU J. Sci. Eng.*, 17, pp. 825–833.

Latarius, K., Schauer, U. and Wisotzki, A. (2019). Near-ice hydrographic data from Seaglider missions in the western Greenland Sea in summer 2014 and 2015, *Earth Syst. Sci. Data*, 11, pp. 895–920.

Leccese, F., Cagnetti, M., Giarnetti, S., Petritoli, E., Luisetto, I., Tuti, S., Đurović-Pejčev, R., Đorđević, T., Tomašević, A. and Bursić, V. (2018). A simple Takagi-Sugeno fuzzy modelling case study for an underwater glider control system, *In Proceedings of the 2018 IEEE International Workshop on Metrology for the Sea; Learning to Measure Sea Health Parameters (MetroSea)*, Bari, Italy, 8–10 October 2018, pp. 262–267.

Loc, M. B., Choi, H., Kim, J. and Yoon, J. (2012). Design and control of an AUV with weight balance, *Oceans — Yeosu, Yeosu*, 2012, pp. 1–8.

Low, K. H. (2006). Locomotion and depth control of robotic fish with modular undulating fins, *International Journal of Automation and Computing*, 4, pp. 348–357.

Naser, F. A. and Rashid, M. T. (2019). Design, modeling, and experimental validation of a concave-shape pectoral fin of labriform-mode swimming robot, *Engineering Reports*, 1(5), pp. 1–17.

Naser, F. A. and Rashid, M. T. (2020). Effect of Reynold number and angle of attack on the hydrodynamic forces generated from a bionic concave pectoral fins, *IOP Conf. Ser.: Mater. Sci. Eng.*, 745, pp. 1–13.

Naser, F. A. and Rashid, M. T. (2020). The influence of concave pectoral fin morphology in the performance of labriform swimming robot, *Iraqi Journal for Electrical and Electronic Engineering*, 16(1), pp. 54–61.

Naser, F. A. and Rashid, M. T. (2021). Design and realization of labriform mode swimming robot based on concave pectoral fins, *Journal of Applied Nonlinear Dynamics*, 10(4), pp. 691–710.

Naser, F. A. and Rashid, M. T. (2021). Enhancement of labriform swimming robot performance based on morphological properties of pectoral fins, *J. Control Autom. Electr. Syst.*, 32, pp. 927–941.

Naser, F. A. and Rashid, M. T. (2021). Implementation of steering process for labriform swimming robot based on differential drive principle, *Journal of Applied Nonlinear Dynamics*, 10(4), pp. 737–753.

Naser, F. A. and Rashid, M. T. (2021). Labriform swimming robot with steering and diving capabilities, *Journal of Intelligent & Robotic Systems*, 103(14), pp. 1–19.

Oo, H. L., Anatolii, S., Naung, Y., Ye, K. Z. and Khaing, Z. M. (2017). Modelling and control of an open-loop stepper motor in Matlab/Simulink, *2017 IEEE Conference of Russian Young Researchers in Electrical and Electronic Engineering (EIConRus)*, St. Petersburg, pp. 869–872.

Ozmen Koca, G., Korkmaz, D., Bal, C., Akpolat, Z. H. and Ay, M. (2016). Implementations of the route planning scenarios for the autonomous robotic fish with the optimized propulsion mechanism, *Meas. J. Int. Meas. Confed.*, 93, pp. 232–242.

Petritoli, E., Leccese, F. and Cagnetti, M. (2019). High accuracy buoyancy for underwater gliders: The uncertainty in the depth control, *Sensors*, 19(8), pp. 1831.

Prestero, T. (2001). Verification of a six-degree of freedom simulation model for the REMUS autonomous underwater vehicle, *M.S. Thesis*, Massachus.

Rashid, M. T. and Rashid, A. T. (2016). Design and implementation of swimming robot based on labriform model, *Al-Sadeq International Conference on Multidisciplinary in IT and Communication Science and Applications (AIC-MITCSA)*, pp. 1–6.

Rashid, M. T., Naser, F. A. and Mjily, A. H. (2020). Autonomous micro-robot like sperm based on piezoelectric actuator, *International Conference on Electrical, Communication, and Computer Engineering (ICECCE)*, pp. 1–6.

Ruiz, S. S., Renault, L. L. and Garau, B. B. (2012). Underwater glider observations and modeling of an abrupt mixing event in the upper ocean, *Geophys. Res. Lett.*, 39, p. L01603.

Seo, D., Gyungnam, J. and Choi, H. (2008). Pitching control simulation of an underwater glider using CFD analysis, *In Proceedings of the Oceans — MTS/IEEE Kobe Techno-Ocean*, Kobe, Japan, 8–11, pp. 1–5.

Singh, Y., Bhattacharyya, S. K. and Idichandy, V. G. (2017). CFD approach to modelling, hydrodynamic analysis and motion characteristics of a laboratory underwater glider with experimental results, *Journal of Ocean Engineering and Science*, 2, pp. 90–119.

Tan, X. (2011). Autonomous robotic fish as mobile sensor platforms: Challenges and potential solutions, *Marine Technology Society Journal*, 45(4), pp. 31–40.

Waldmann, C., Kausche, A. and Iversen, M. MOTH-An underwater glider design study carried out as part of the HGF alliance ROBEX, *In Proceedings of the 2014 IEEE/OES Autonomous Underwater Vehicles (AUV)*, Oxford, UK, 25–29 July 2014, pp. 1–3.

Wood, S. (2013). State of Technology in Autonomous Underwater Gliders, *Mar. Technol Soc. J.*, 47(5), pp. 84–96.

Yu, J., Zhang, F., Zhang, A., Jin, W. and Tian, Y. (2013). Motion parameter optimization and sensor scheduling for the sea-wing underwater glider, *IEEE J. Oceanic Eng.*, 38, pp. 243–254.

Zhang, F., Thon, J., Thon, C. and Tan, X. (2012). Miniature underwater glider: Design, modeling, and experimental results, *IEEE International Conference on Robotics and Automation RiverCentre*, Saint Paul, Minnesota, USA May 14–18.

Chapter 6

CONCLUSIONS

This book has presented a study of Labriform swimming robot design that is used with a pair of pectoral fins only in swimming.* This study started by selecting the optimal shape morphology that produces the highest thrust. Three different fins were tested throughout this work. The results showed that the lowest aspect ratio fin (Fin1) produces the highest thrust at high velocities but requires higher energy due to its large surface area which consequently generates a high drag. On the other hand, an intermediate aspect ratio fin (Fin 2), gives an acceptable thrust at different velocities, whereas the highest aspect ratio fin (Fin3) generates the lowest thrust and it was excluded in this work. For that reason, Fin1 was chosen in the diving control mechanism.

In Chapter 3, the complete design of the robot has been proposed with an elliptical cross-sectional area that minimizes the water pressure during swimming to a minimum. Five different power to recovery stroke ratios were simulated and validated experimentally. The results showed that the maximum thrust obtained in the moderate ratio of R_F is 3:1. This ratio is chosen to be the optimum ratio since it produces the largest thrust during the power stroke

---

*Despite the engineering complexity of the various robots, the gallery of the presented experimental trends allows the model identification of easy mathematical models. The reader interested to perform numerical works could refer directly to Mofeed Turky Rashid (mofeed.rashid@uobasrah.edu.iq), Electrical Engineering Department, University of Basrah, Basrah, Iraq.

and the lowest drag at recovery stroke time. Several simulations and practical experiments are conducted, including:

(1) The open-loop response of both position and velocity of the fin motion showed that a noticeable ripple may affect the overall performance of the system performance, therefore, it should implement a complete system controlling mechanism via a closed-loop PID controller to adjust the movement of the pectoral fins and reduce ripple in position and velocity signals.

(2) Study the effect of the starting angle of the fin, where the results showed that the maximum angle to start pushing in the power stroke phase is at 50°. The maximum range is limited to 100°.

(3) The efficiency of the proposed design is tested and compared to some real fish numbers. Strouhal number (St) is calculated and the results showed that it is relatively far from the real extent range of the fish, due to the reliability of the proposed design in swimming on pectoral fins only, and this leads to obtaining a relatively small amount of forwarding velocity, which negatively affects the values of St.

(4) Reynolds number is dealt with as another factor in the proposed design. It has been tested at different starting angle values and ranges are explored of the effect of differing Reynolds numbers on the drag force generated at each angle. The results showed that the minimum drag forces are obtained at 50° of starting angle even when increasing Reynolds number.

(5) The amplitude to body length ratio (A/BL) was studied and compared to the true ranges of fish. The results showed very close results to the biological ranges, especially when increasing the oscillation amplitude to 100°.

To implement the steering process, two steering systems have been considered, the first one depends on the caudal fin of boxfish, in which turning torque in the flow basin is measured using a physical model with attachable caudal fin closed or open at the different body and tail angles, and different water flow velocities, as the results are evaluated by the simulated computational fluid dynamics (CFD), which indicates that the caudal fin achieves steering. The results show that the effective change of caudal fin shape and direction plays an important role in controlling steering torque in boxfish. The caudal fin is both a track anchor and a rudder. These results correlate with the swimming behavior of boxfish in their natural environment. Although this new information is an important first step in improving our understanding of controlling steering motions in boxfishes, further study is needed to reveal how all components of the boxfish thickness motor system work together, from a dynamic perspective, during lateral storm flows and transformation, while, the other system is based on the differential drive principle of the two-wheel mobile robot is adopted. The results showed a good performance in minimizing turning radius while using pectoral fins only without any intervention from using caudal fin, it gives a competitive output in comparison with other swimming robots in literature. The minimum turning radius achieved, about 0.40 of body length for a turning period of about 15 sec to complete one cycle.

Last but not least, the buoyancy control mechanism is investigated carefully by two systems as presented in Chapter 5. In the first system, the mechatronic design and fabrication of a prototype Carangiform (i-RoF) swimming robot with a two-link propulsion tail mechanism was presented. The prototype consisted of three main parts: a front rigid body, a two-link thruster tail mechanism, and a flexible tail fin. A biomimetic motion control structure based on a central pattern generator (CPG) is proposed to simulate the vigorous, fish-like swimming gait. While the center of gravity (CoG) control mechanism is designed and placed in the front rigid body to ensure 3D swimming ability. Therefore, important key issues related to 3D swimming capabilities and biomimetic design are emphasized and detailed in this study. A solid torpedo-shaped front body is designed to contain the electronics, sensors, and CoG control mechanism. The CoG control mechanism successfully provides buoyancy and diving movement capabilities. Motion control is adapted based on a central pattern generator (CPG) to generate strong, smooth, and regular oscillating swimming patterns, whereas the second system is the mass shifter approach, as a glider model.

The glider swimming robot has two types of independent actuators: the first one is to represent the center of gravity, and the second one used for pectoral fin actuation. As the pitch motor starts to rotate from its center of gravity position toward the other end of the lead screw, the movable mass will consequently pitch the robot down with about 40° and the designed robot is able to swim for about 30 cm underwater depth. The simulation and practical experimental results showed the ability of the robot to dive and float following a predetermined sawtooth pattern.

Index

www.ingramcontent.com/pod-product-compliance
Lightning Source LLC
Chambersburg PA
CBHW050603190326
41458CB00007B/2157